Evolving Global Information Infrastructure and Information Transfer

Evolving Global Information Infrastructure and Information Transfer

ROBERT J. GROVER,
ROGER C. GREER,
HERBERT K. ACHLEITNER,
and KELLY VISNAK

LIBRARIES UNLIMITED

AN IMPRINT OF ABC-CLIO, LLC
Santa Barbara, California • Denver, Colorado • Oxford, England

Library of Congress Cataloging-in-Publication Data

Grover, Robert, 1942–
    Evolving global information infrastructure and information transfer / Robert J. Grover, Roger C. Greer, Herbert K. Achleitner, and Kelly Visnak.
        pages cm
    Includes bibliographical references and index.
    ISBN 978-1-61069-957-0 (pbk : alk. paper) — ISBN 978-1-61069-958-7 (ebook)
1. Information superhighway.   2. Information science—Social aspects.
3. Information society.   4. Information technology.   5. Information services—Technological innovations.   6. Electronic information resources.   7. Digital libraries.   8. Libraries—Forecasting.   I. Greer, Roger C., 1928–2014
II. Achleitner, Herbert K.   III. Visnak, Kelly.   IV. Title.
    ZA3225.G76 2015
    303.48'33—dc23   2014046767

ISBN: 978-1-61069-957-0
EISBN: 978-1-61069-958-7

19 18 17 16 15   1 2 3 4 5

This book is also available on the World Wide Web as an eBook.
Visit www.abc-clio.com for details.

Libraries Unlimited
An Imprint of ABC-CLIO, LLC

ABC-CLIO, LLC
130 Cremona Drive, P.O. Box 1911
Santa Barbara, California 93116-1911

This book is printed on acid-free paper (∞)
Manufactured in the United States of America

Contents

In Memoriam

Roger Greer, our coauthor, mentor, and close friend, died one week after we submitted this manuscript to our acquisitions editor. Although Roger's health had deteriorated in recent years, his mind was as sharp and creative as ever. Likewise, his sense of humor was intact until the end.

During the two years that we wrote this book, Roger was fully engaged. We met via conference call every Tuesday morning during those two years to discuss progress and problems encountered. We also met at his house over a long weekend twice a year to review our progress and to shape the direction, content, and overall philosophical direction of the book. He was our host, and at meal times he guided us to one of his carefully selected restaurants that served pork chops and mashed potatoes.

Our weekend retreats and weekly conference calls contributed to this book, but we also covered such topics as family, politics, health issues, annoyances, and each other's foibles. The conversations always were a highlight of our week, and Roger's wit and incisive comments were a significant part of each conversation. Our time together was an intensive intellectual experience and reminded us of all those years of time spent with Roger in his office, where he proposed his latest theoretical models and challenged us to respond to, critique, and add to his ideas.

Roger was an oversized presence both physically and intellectually. He was six feet two inches in height, and his creativity, knowledge, foresight, wit, and his impact on our careers and lives, were likewise imposing. He was more like a big brother to us than a colleague; he was a member of our family. We'll miss him, and he'll always be a part of us.

We dedicate this book in Roger's memory.

Bob Grover
Herbert Achleitner
Kelly Visnak

Preface

We began this writing project as an effort to fulfill a need for a book that would make sense of the swirling array of changing technologies and information formats that confront us in this digital age. More than that, we wanted to address the bigger picture of a changing global society that is facing new challenges never before experienced. We wanted to provide a perspective on how society and technology have evolved over time to arrive at our current state in this digital age, and we wanted to apply models that would help us explain the technologies, organizations, and people that form the global information infrastructure.

As a result of our experiences in the library and information professions, our research, our reading, our conversations with others, and our discussions together, we believe that this book offers a fresh approach for analyzing and explaining the complex flow of information in today's world.

We also wanted to explore the library and information profession's transition from the bibliographic (book- and materials-centered) paradigm to a new digital paradigm that is client-centered. We wanted to address the changing role of library and information professionals, from their emphasis on information and technology to an emphasis on relationships, as they assume a greater role in helping their clientele derive meaning from the vast amount of information that engulfs us. The transformed library and information professions must transition to a dynamic leadership role that requires a deep understanding of the information infrastructure.

Library and information professionals must be aware of the forces that influence change in order to respond and lead in a profession that is central and vital in contemporary society. Instead of focusing on technologies, which are always changing, this book looks at the "big picture" of how information is created, recorded, mass-produced, distributed, diffused, and utilized in society. This unique approach will enable readers to understand better the complexities of today's world and to assume a leadership role.

Acknowledging and building upon the transportation, communication, and information systems established before the infusion of digital technology, we examine the global infrastructure that supports the creation, dissemination, diffusion, and utilization of information. While the transportation infrastructure is an example of a recognized and vital support for society, the information infrastructure, while increasingly important, has not received similar recognition or scrutiny. Furthermore, the concept of "information infrastructure" is usually confined to the technology and digital components of storage, retrieval, dissemination, and preservation of information. This book expands this concept to include information transfer and all of its components—creation, recording, production, dissemination, organization, diffusion, utilization, deletion, and preservation of information. This information transfer model reveals the roles of individuals (professionals and paraprofessionals) and organizations at each stage.

This examination of the information infrastructure in the context of information transfer provides a framework for looking at a complex and emergent digital world that is far different from the analog world we experienced before the widespread use of computers and the Internet.

We examine each of the stages or processes of information transfer; however, the focus is on diffusion and utilization of information, in order to underscore these processes and their critical role in the development and furthering of a society's culture. Unless a society has a vibrant information infrastructure, it cannot progress economically, politically, intellectually, or culturally.

Our author team has worked together for many years, and in this book we have applied many of the models that we have developed or studied through our research, writing, and teaching. Over the years we have discussed and expanded our models and theories of information transfer, and we share them with the reader in this book.

We are grateful to our professional colleagues for their openness and willingness to share their ideas with us; we are especially grateful to Jackie Lakin, Information Management Consultant, Kansas State Department of Education; Sharon Morris, Director of Library Development, Colorado State Library; and John Sheridan, Dean of Libraries, Emporia State University for their time and ideas as we discussed current trends. We are grateful to Daniel Brogren and Nancy Marlatt for sharing their graphic talents as they helped us create graphic models that appear in chapters 5 and 10.

We are thankful for our families and friends who supported us when we spent weekends away from home with obligations to meet with our author team to discuss and write.

And we express our gratitude to our acquisitions editor, Blanche Woolls. Her advocacy and encouragement made this book possible. Thank you, Blanche!

1

Introduction to the Book

Chapter Overview

"Complexity" is a term that best characterizes today's fast-paced, changing world. Our culture, our values, our family structure, our educational system, our cars, the way we communicate, the way we live—nearly every aspect of our lives—has radically changed in the last 20 years. Another way of expressing it is that this generation is facing a much different world than previous generations.

The driving force for these changes is information and communication technology, and in this book we provide an analysis and examples of our current society and how information technology has changed dramatically the way we create new information and knowledge, how we communicate it, learn from it, use it, and store it. The dynamic, nonlinear flow and our use of information have changed our values and belief systems and the context in which we operate.

These changes have created a world that is fast-paced and confusing. To address the complex world swirling about us, we will explore our culture, technologies, and flow of information to make sense of our often confusing information infrastructure.

In this chapter, we introduce an overview of our society today, using some models that provide an understanding of the many forces that have converged to create this complex milieu. We provide a perspective of the information infrastructure and the concomitant role of information transfer that is the driving force in shaping the content of our culture. We also define terms used frequently and describe the purpose of the book and the audience for whom it is intended.

Introduction to the Information Infrastructure

We do things differently today than we did before, and at the heart is the lifeblood of our culture: information and knowledge. Information is to our society what blood is to our bodies, and the vascular system that carries our blood is like the information infrastructure that supports the creation and other functions of information that result in its effective utilization.

In an earlier work, we defined the "information infrastructure" in broad terms:

> The information infrastructure is a **global** network of **people, organizations, agencies, policies, legislation, processes, and technologies** organized in a loosely coordinated system to enhance the creation, production, dissemination, organization, diffusion, storage, retrieval, and preservation of information and knowledge for people. The primary objective of this network is the diffusion of knowledge for a society. (Greer, Grover, and Fowler 2013, 100)

Although technology is an important component of the information infrastructure, we wish to underscore the importance of the human element: the information users and the personnel who design and implement the various components of this vast infrastructure. In our discussion of the infrastructure, we also want to underscore the influence of information and knowledge on a society's culture; hence, the information infrastructure's impact on culture is significant. We also recognize that culture, along with other external forces, impacts the information infrastructure. As we examine the organizations and systems that are components of the information infrastructure, we will evaluate the systems and services according to their effectiveness in addressing the information needs of users.

This infrastructure is like the vascular system in the human body; it provides the system that supports the transfer of information and knowledge to a society, supporting its culture. This system is vital to the life of the culture, just as the vascular system is critical to the body. If there is a failure in the system (like a blood clot in the body), the infrastructure and information transfer system is compromised, with potential harm to the entire system and the body or society that it supports.

A key concept in our study of information infrastructure is "information transfer." In our previous work, we have defined information transfer as "the communication of a recorded message from one human or human mind to another" (Greer, Grover, and Fowler 2013, 59). Although the sender and receiver of a message in typical communication may be contemporaries, information transfer requires a recorded message, and the sender and receiver may not be contemporaries. Stated another way, information transfer is asynchronous.

The stages or processes of information transfer are the following: creation, recording, mass production, dissemination, bibliographic control, organization by discipline, diffusion, utilization, preservation, and discarding. This model for understanding information transfer is the model we use for examining the information infrastructure.

Recent advancements in technology have provided new methods for the transfer of information. For example, Snapchat combines visual and textual information as well as highlights and layers information for easier comprehension, focusing simultaneously on the sender and receiver of the message. Unlike a book, which appears flat and is designed for the reader to progress in linear fashion, the above example illustrates nonlinearity and the dynamic interaction of information in various formats enhanced by coloring and free-style design.

Thus, technology enables us to break through the barriers of distance and time to work together, solving problems, and to learn, as we never have before. The Internet enables instant transmission of information in a variety of formats to numerous sites and people, enabling a level of participation impossible until this digital age. Course software combined with blogs,

Figure 1.1 Evolution of Information Transfer

Twitter and other social media, and other forms of communication is another example of technology revolutionizing education. Information transfer is explored in more detail in chapters 4 and 5.

As a result of this new way of communicating and transmitting information, the learning process has changed. In other words, the diffusion of information has been speeded up, and the diffusion to mass audiences very quickly has transformed education, from preschool through graduate education and beyond. An example of this transformation is the introduction of course software into distance education, whereby members of a class may span several states (or continents) yet be connected via asynchronous exchanges of ideas on course discussion groups and via synchronous communication through audio and video conferencing. Training of sales staff by insurance companies and the continuing education of attorneys have been changed to enable learning from experts located at sites distant from the learners. Universities are offering courses to students thousands of miles distant. Video conferencing, texting, and Twitter feeds enable a faculty member and students to discuss readings and share experiences in real time.

The transmission of knowledge is a critical element in the evolution of a culture; consequently, the infrastructure that enables transmission of that knowledge is a key factor in supporting cultural growth. It is our observation that the information and knowledge infrastructure—that is, the loosely coupled systems of creation, dissemination, organization, diffusion, and utilization—has been ignored in discussions of our contemporary society; instead, attention has been directed to the information and communication technology advances, without corresponding analysis of the underlying subsystems of information transfer.

This book provides an explanation of the rapidly changing, complex flow of information in the context of culture, policy, technology, and economics. The influence of society and culture on information transfer is examined in Chapter 3. We examine the information and knowledge infrastructure: those human and technological networks that support information creation, production, dissemination, and diffusion in the promotion of learning and effective use of information in society.

In so doing, we recognize the evolution of information in society from the time before the invention of the printing press, which we designate "Information Transfer 1.0." We designate the age of print from Gutenberg (ca. 1450) until the widespread use of the Internet (ca. 1990) as "Information Transfer 2.0"; the digital age that actually began with the invention of the personal computer and spawned widespread use of the Internet, we call "Information Transfer 3.0" (see Figure 1.1).

Understanding the Digital Age

The above timeline correlates with the general belief in Western society that the advancement of civilization is bound to the advancement of knowledge. This belief is based on the powerful notion of "progress" through scientific achievements. Humankind has been on this journey since the

Renaissance—a journey of modernization and the betterment of the human condition. Progress was the result of a series of scientific revolutions or paradigm shifts. Thomas Kuhn (1970) explained how scientists work and change their viewpoint, resulting in a paradigm shift. Schwartz and Ogilvy (1979) traced the evolution of theories in disciplines and how we change our belief systems, providing a list of characteristics that describe the emergent paradigm. Capra (1996) took an ecological view to let us see societal systems through a web of relationships, which helps in conceptualizing information transfer from a systems perspective. The concept of paradigm shifts in society is explored in Chapter 2.

It is the authors' belief that the system's perspective is the most viable model in understanding and navigating the information transfer processes and environmental context. We trace the evolution of information transfer from the Gutenberg world to the digital world, from a mechanistic perspective to a systems perspective, in explaining the evolution of the information transfer systems.

An Emerging Participatory Culture

The Internet provides an ideal space for a participatory culture by simply providing information freely, in an open and widely distributed way. Online stories and video games have moved people into a culture of participation. Readers who unite through the Internet participate in solving a riddle and then gain access to the next chapter in the book; they are involved in a new and collaborative way of using information.

A participatory culture can be explained as an informal community that gathers together to support an activity. Elements of the participatory culture can be found in a variety of activities that may involve learning, creation, community life, or democratic citizenship. Storytelling is one such activity. Readers interested in fictional stories can go to RinkWorks, where a different type of eBook, known as a "Story Hunt," can be found. The first chapter of each Story Hunt is made openly available on the Internet. The reader gains access to each subsequent chapter by relying on clues from the chapter to answer a riddle.

Some riddles have been so challenging that readers have turned to social media to work with other people who are also reading the same story and working to find a solution. These readers may range from your neighbor next door to those from across town or across distant geographic borders. The social media that readers turn to as they work together on a Story Hunt includes forum visits or hint boards that unite those specifically interested in working collaboratively to solve the riddle. This participatory culture is an example of the many changes we face as we study our information infrastructure and the social forces that interact with it.

Purpose of This Book

The purpose of this book is to provide clarity to our rapidly changing, complex society as it is bombarded with new information and technologies. By applying existing models of information transfer, we will help the reader understand how information is created, disseminated, and used, explaining how components of the information infrastructure influence culture and our understanding of the world. We examine the information infrastructure so that readers will be able to understand their roles in the information transfer process; librarians and information professionals will see their professional roles, especially related to the diffusion and utilization of information and knowledge through policy development. This book addresses the need for a holistic analysis of the role of library and information professionals and organizations in a complex and rapidly changing information infrastructure.

A review of pertinent and current literature indicates that little attention has been directed to the role of information transfer within the information infrastructure as a building block of a culture. The purpose of this book is to accept this challenge. In doing so, we focus on the diffusion of information and its utilization.

In this book we examine the information infrastructure as it has evolved from print to the current digital world. The successful evolution of a culture depends on the role of information and its modes of information transfer among the constituents of that culture. Our thesis is that the quality of the information infrastructure determines the impact of the information transfer processes; that is, the creation, reproduction, dissemination, diffusion, and utilization of information. In the process of examining this infrastructure, we also assess the effectiveness of the various processes and suggest improvements.

Unless we have a vibrant information infrastructure, our society cannot progress economically, politically, or culturally. Currently, the concept of information infrastructure is confined to the technology and digital components of storage, retrieval, dissemination, and preservation of information. This book expands that concept to include information transfer and all of its components—the creation, recording, production, dissemination, organization, diffusion, utilization, preservation, and deletion of information. As blood depends on a vascular system to support life, information relies on an infrastructure to support the life of a culture; this book examines the information infrastructure and its significance as a cultural and economic force in our rapidly changing technological society.

In other words, we are preparing the framework for studying and coming to terms with a complex, confusing, and emergent digital world. In so doing, we are providing *an* answer to the understanding of our information and knowledge society, not *the only* answer. We are challenging the reader with a wide-angle lens for studying in a comprehensive manner our information infrastructure. We are also empowering the reader to understand how technology, through the Internet and social media, has changed communication patterns and our 21st-century culture.

When we depart from the linearity of the pre-digital age (Information Transfer 1.0 or 2.0), technology encourages new ways of creating, disseminating, and utilizing information. The traditional single author gives way to multiple voices, fast change, and lack of control. The World Wide Web strips the world of control, as evidenced by some governments attempting unsuccessfully to stop the exchange of ideas through social media. In the past, the goal was to increase information transfer (e.g., Gutenberg's printing press) so that more voices could be heard. Now the dissemination of information is happening a thousandfold, as smart phones provide access to the Internet and an individual has an audience worldwide.

Intended Audience

The intended audience is library and information professionals and students, as well as educated adult readers with an interest in the social implications of the digital age. This audience includes librarians in all types of settings, information specialists of all types, the research community, educators, information workers in public and private sector organizations, and policy makers. However, we also intend these ideas for the general reader who wants to understand better the technology and information resources available in dizzying quantities, presenting a challenge to his or her understanding. The book is of interest to undergraduate and graduate students in information studies and sociology, especially the sociology of knowledge.

Definition of Basic Terms

In this book we frequently use the terms "information" and "knowledge," words that often are used interchangeably. It is important to distinguish the differences between these terms, but to do so we must also look at the terms "data" and "wisdom" as we defined them in our earlier work, *Introduction to the Library and Information Professions* (Greer, Grover, and Fowler 2013). We base our definitions on Harlan Cleveland's important book *The Knowledge Executive* (1985).

Data

Data are the rough materials from which information and knowledge are formed; that is, undigested observations, or unvarnished facts, as Cleveland calls them (1985, 22). For example, researchers collect data from interviews, observations, surveys, and other means in order to analyze it for research purposes. Researchers are not the only people who collect data on a regular basis; all of us are bombarded with data all the time. To make sense of this barrage, data must be synthesized. Data may take the form of words, numbers, or visual images. By themselves, data make no sense. For example, these data alone have no meaning unless placed in a context, as noted below: red, 4, 2, and 8.

Information

Using Cleveland's (ibid.) definition, information is organized data. Given connections or context, data can form information. In this case, we may be watching a baseball game, and the Red Sox are leading their opponent four runs to two in the eighth inning. Now we have information, because we have put the numbers together in a meaningful way.

Knowledge

Knowledge, according to Cleveland, "is organized information, internalized by me, integrated with everything else I know from experience or study or intuition, and therefore useful in guiding my life and work" (ibid.). The noise of a distant train's whistle, a passing car, conversation in the next room, the sound of a furnace or air conditioner, or the television in the background as we read all give us data or information that we may reject or do not retain or remember. When we watch the television news, read a newspaper, or read our e-mail, we remember only a small percentage of what we read, view, or listen to. That which we remember or incorporate in our memory becomes our personal knowledge. In other words, information that is processed, selected, and synthesized by a human becomes knowledge.

Knowledge is also processed by groups of people, and we will refer to social knowledge when we discuss information transfer and knowledge diffusion. As with personal knowledge, social knowledge requires analysis, selection, and synthesis of information in order to be accepted as knowledge. We discuss this process in much more detail later in chapters 6 and 7.

Wisdom

When knowledge is integrated into the thinking of human beings and incorporated into their decision-making processes, it becomes wisdom. Cleveland defines wisdom as follows:

> Wisdom is integrated knowledge, information made super-useful by theory, which relates bits and field of knowledge to each other, which in turn enables me to use the knowledge to do something. (1985, 23)

Using the example above of the baseball game, we know that the Red Sox are leading their opponent 4–2 in the eighth inning. This information tells the fan important game information, and combined with the individual's experience of following baseball games over time, the fan knows that the Red Sox do not have a comfortable lead and could still lose in the ninth inning. Leaving the game before the ninth inning could mean missing the last inning and any action to come in that inning. In other words, new information is taken in, evaluated, and acted upon based upon the existing store of accumulated knowledge or wisdom. Wisdom, then, is the accumulation of a lifetime's experience and testing of knowledge. An understanding of data, information, and knowledge are basic to an understanding of the information transfer process, which provides the framework for studying the information infrastructure.

Information Transfer

"Information transfer" is a type of communication, as described above. It can be defined as the communication of a recorded message from one human to another. While communication assumes that the sender and receiver(s) of a message are contemporaries, information transfer requires a recorded message transmitted through a medium that enables senders to transmit ideas to people who are not their contemporaries. In other words, information transfer is asynchronous. We use the information transfer processes as a framework for studying the information infrastructure. See Chapter 4 for a more detailed description of information transfer.

Summary

The driving force for the fast and all-encompassing change in our society is technology. This book examines current society and how technology has changed dramatically the way we create, disseminate, diffuse, and utilize new information and knowledge. These changes have resulted in a participatory culture that has changed our lives. By studying the grand theories of Thomas Kuhn and others, we learn that cultural changes are accompanied by paradigmatic change as well, changing the way we view the world.

In this book, we offer the framework for understanding our complex, confusing, and emergent digital world. We provide the reader with a lens for examining, in a comprehensive manner, our information infrastructure. We also empower the reader to understand how technology, through the Internet and social media, has changed communication patterns and our 21st-century culture. In the next chapter, we examine more closely this culture as a foundation for our examination of the information infrastructure.

References

Capra, Fritjof. 1996. *The Web of Life: A New Scientific Understanding of Living Systems*. New York: Anchor Books.

Cleveland, Harlan. 1985. *The Knowledge Executive: Leadership in an Information Society*. New York: Truman Talley Books/E. P. Dutton.

Greer, Roger C., Robert J. Grover, and Susan G. Fowler. 2013. *Introduction to the Library and Information Professions*. 2nd ed. Santa Barbara, CA: Libraries Unlimited.

Kuhn, Thomas S. 1970. *The Structure of Scientific Revolutions*. Chicago: University of Chicago Press.

Schwartz, Peter, and James Ogilvy. 1979. *The Emergent Paradigm: Changing Patterns of Thought and Belief*. Report issued by the Values and Lifestyle Program, April.

2

Contemporary Society

Chapter Overview

Change is a constant in our world of the 21st century, and technological advances fuel that change. In this chapter we examine models that explain the fundamental changes that have rocked Western society in the last 600 years. We also describe the environmental factors that influence our perception of the world and impact the creation, dissemination, diffusion, and utilization of information and knowledge.

Historical Roots of Contemporary Society

In 1439 in Mainz, Germany, Johann Gutenberg invented the printing press and changed virtually overnight the way we communicate, learn, see, and do things. The power of the Catholic Church was considerably reduced; Latin as the lingua franca was replaced by national languages. Print allowed languages to become fixed, standardized, easily spread, and preserved. Handwritten manuscripts could be lost forever, the same as with carved wooden blocks. However, with the printing press, the chances of a copy surviving were much greater.

Printing brought a fundamental change in the advance of communication, breaking the monolithic power of government and church and weakening their ability to control. It brought about a beginning of national identities and recognition of the power of knowledge. It also fundamentally changed the world of economy, driving new commercial interests and bringing about the rise of capitalism. It was a movement away from the inward-looking medieval world toward a more open, outward-looking perspective. More importantly, this new technology moved access to knowledge from random and limited access in archives to the beginning of libraries organized to increase access and to diffuse and use knowledge. A brief history of other technologies that changed communication and information transfer patterns is found in Chapter 3.

Revolutionary Change in Contemporary Society

Five hundred years after the invention of the printing press, the world is again radically changing. Just as the Gutenberg invention significantly increased the production and dissemination of information, so does today's networked world. Not only the spread of knowledge, but also its diffusion and utilization have become a political and economic necessity.

What has changed? What is this emergent world that surrounds us and that we speak and speculate about every day? For one, it is a networked world with an ever-increasing transborder flow of goods, ideas, and people: an economy in which a significant portion of the Fortune 500 companies listed 30 years ago have completely disappeared, merged and lost their identity, or have become smaller and less significant. We have become a world where the once dominant Western values are being questioned and challenged, where conventional warfare is being replaced by individual terrorists and cyber warfare, and where soldiers on the field are being replaced by remote devices piloted from a great distance with devastating effect.

The hallmark of the networked world is connectivity and instant communication. Digitization, miniaturization, and the merging of a bundle of technologies, like the cell phone, for example, are just the beginning of a reconceptualization of not only how we live, but also how we work and think. The challenge we all face is to acquire the values of the networked world that requires a different way of thinking and doing, a world marked by complexity, diversity, and individuality, where information and information technology provide the potential for every person and every voice to be heard. That is the cultural revolution taking place today.

An Analysis of Our Current Paradigm Shift

The creation, production, diffusion, and utilization of knowledge are central to the advancement of humanity. The Greeks introduced the notion of salvation through knowledge. Thomas Kuhn posed the fundamental question: How did we move from Aristotle's physics and concept of the universe to those of Einstein? The result of this inquiry led to his remarkable book *The Structure of Scientific Revolutions* (1970), where Kuhn developed the notion that scientists work within a distinctive paradigm. Kuhn describes "paradigm" in a variety of ways, but simply put, a paradigm is a set of fundamental beliefs, assumptions that provide a coherent worldview; that is, the reality we perceive and how the world works. It can be best described as theories, concepts, values, methods, and perspectives.

Paradigms give us a general perspective of the world and a way of coping with the complexities of life. They provide us with an intellectual framework: what research should we pursue, what is true, real, important, legitimate, and reasonable. It is the lens through which we see a reality.

Fundamentally Kuhn analyzes how scientific revolutions occur. He proposes a paradigm shift sequence: existing paradigm, normal science, anomalies, crisis stage, revolutionary stage, and new paradigm (see Figure 2.1). Normal science is research that builds on past scientific achievements and is accepted by a scientific community and directs its practice, using the same rules and research standards. During the normal science stage, a consensus exists that guides research in a scientific community. Normal science is a puzzle-solving activity. The crisis stage occurs when anomalies are produced, often inadvertently, and the scientific community

Figure 2.1   Thomas Kuhn's Conception of a Paradigm Shift

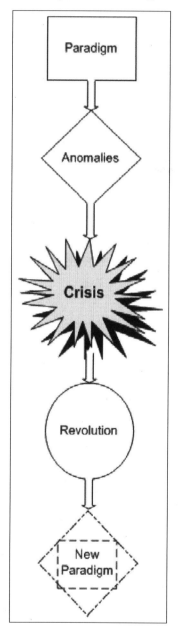

realizes they violate expectations posed by the existing paradigm. This eventually leads the community to a new set of practices and commitments; a reconstruction of the field takes place.

Most importantly, as a new paradigm emerges, a different view of the nature of reality emerges. New research questions, methods, and standards emerge; the results are a reconceptualization of the field. The new paradigm, according to Kuhn, is "incommensurable" with the old paradigm.

For example, in the world of print from the scientific age to the information age, our belief system seemed to be intact. Our culture appeared to be integrated, holistic, directed, and consistent. However, when belief systems are challenged, cultures become fragmented or disintegrated, lose direction, and are seemingly at odds within. Our sense of how the world works and our place in it is challenged. Knowledge, once privileged and integrated, loses its monopoly, as demonstrated by Wikipedia, where information is created by a network of anonymous authors, rather than by individual experts.

The new view of reality started to emerge at the beginning of the 20th century, in physics. Albert Einstein's theory of relativity, Max Planck's quantum theory, and Heisenberg's uncertainty principle destroyed the Newtonian worldview of a mechanistic world. Uncertainty, probability, and relativism destroyed the neat Newtonian perspective of a harmonious universe.

The destruction of the mechanistic and deterministic world is analyzed in an excellent report by Peter Schwartz and James Ogilvy, *The Emergent Paradigm: Changing Patterns of Thought and Belief* (1979). They studied a number of disciplines and discipline-like fields and their contribution to the "emergent paradigm." A review of physics, chemistry, mathematics, philosophy, politics, psychology, linguistics, religions, and art, as well as discipline-like areas, such as brain theory, ecology, evolution, and consciousness, led to our understanding of the emergent paradigm and proposed a list of characteristics:

1. From simple toward complex views of phenomena

2. From hierarchic toward heterarchic views of order

3. From mechanical toward holographic metaphors

4. From a view that the unknown can be determined toward an acceptance of indeterminacy

5. From linear cause and effect toward mutual causality

6. From a view of change as a planned assembly of events toward a sense that change is ongoing and spontaneous

7. From a belief in objective research toward an acceptance of perspective research (Achleitner and Hale, 1988)

The old paradigm's mechanistic value system championed order, domination, and control. What emerges is a new view of reality where relationships are weblike, dynamic, and nonlinear. Complexity, diversity, interaction, and open systems are the new reality. The question then arises, how do we navigate in a world of open systems, dynamically linked where change occurs unpredictably, as exemplified every nanosecond on the Internet?

While physics destroyed the Newtonian worldview, biology since Darwin gave us a biological systems perspective in which systems evolve through complex, mutual causal processes. Open subsystems, like the information transfer subsystems in the digital age, are capable of sudden evolution into new relationships. For example, incapable of maintaining a stable equilibrium as a closed system, the former Soviet regime, when faced with an external or internal challenge, such as Western economic productivity and growth or the citizens' resistance to the communist system, collapsed rather suddenly.

As Capra (1996) points out, evolutionary thinking is the science of complexity—of change, growth, and development. From this point of view, knowledge is not linear but instead "mutually causal," not a planned assembly of data and facts built on previous knowledge, but rather shaped by positive feedback and feedforward (Schwartz and Ogilvy). Science is actually an interconnected network of relationships with no firm grounding.

This biological/ecological thinking of systems dynamically interacting, forever changing, and in a constant "emergent" state is how we view Information Transfer 3.0. In the Kuhnian sense, the digital age has created revolutionary conditions in how knowledge is created, produced, organized, used, preserved, and destroyed. Therefore, the systems perspective is the best lens and framework for understanding the information technology revolution.

Characteristics of the Emergent Paradigm

A lens into understanding the networked digital world is the work of Peter Schwartz and James Ogilvy (1979) and their analysis of theory building in different disciplines and the "emergent" characteristics of the new paradigm. What they describe is the collapse of certainty in society, as evident in the intellectual revolutions taking place in the sciences, in the paradigms of fields such as physics, chemistry, and biology but also in the humanities and social sciences. An analysis of the shift between dominant and emergent paradigms is displayed in Table 2.1. Each of the elements of the table is then described.

Simple to Complex

The emergent paradigm recognizes the complexity of contemporary society. As a result, simple solutions have given way to complexity. Even the marketplace is much more complex. For example, the "Big Three" automakers (General Motors, Ford, and Chrysler) have given way to a multitude of international manufacturers that often rival or surpass American automakers in quality. In addition to the expanded market of autos from France, Great Britain, Germany, Japan, Korea, and China, among others, each automaker provides a plethora of models, colors, and options to choose from. To further complicate the auto market, it is increasingly difficult to identify the nationality of the manufacturer. Until the 1990s, American cars were made with parts made in the United States.

Similarly, life has become more complex. New technologies provide many options for entertainment and for information sources. News is available to us in printed and electronic newspapers and magazines, by television or radio, from the Internet via telephone or iPod, and through informal

Table 2.1   Characteristics of a Paradigm Shift

Dominant Paradigm	Emergent Paradigm
Simple	Complex
Hierarchy	Heterarchy
Mechanical	Holographic
Determinate	Indeterminate
Linearly causal	Mutually causal
Assembly	Morphogenesis
Objective	Perspective

networks that have been extended through social media. Furthermore, the complexity is exasperated by the rate of change. New inventions are invading the marketplace and being adopted at a rapid pace. This pace causes frustration, and some people throw up their hands, exclaiming that they have given up on keeping up.

Hierarchy to Heterarchy

The structure of organizations has changed from a traditional hierarchical structure like that of the military with generals, colonels, majors, captains, lieutenants, and noncommissioned officers to a flattened heterarchy. A hierarchy has specified levels of authority, with communication from top down, or from bottom up, in a specified linear pattern. The communication is expected to follow the levels of authority one by one.

A heterarchy, by contrast, is more fluid and more casual in organization and communication. A marriage relationship is a heterarchy when the couple negotiates who is in charge, and that may vary according to the situation. Currently, an American marriage is likely to be a partnership, not a dominating relationship. In families, parents do not rely on their status as parents for authority; they must reason with their children.

In organizations, a heterarchical or flattened structure enables people at any point in the organization to have access to decision makers at any level. Heterarchy is the new paradigm organization plan in organizations; the top management is likely to be more interested in the content of employees' heads than the use of their hands. One of the characteristics of cell phones, e-mail, texting, and instant messaging is that any member of the organization can be reached more easily at any time by any other member of the organization. As a result, the hierarchy is deflated and replaced by a heterarchy. The inflexibility of communication by level in the hierarchical organization is replaced by flexibility in the communication system, making the organization more fluid and responsive to the need for change.

Similarly, library and information systems like indexing can be made more flexible and responsive to the needs of clientele by using natural language searching or by creating multiple access points using many indexing terms. The former limitations of paper information sources and their consequent necessity to limit indexing terms have been replaced by the use of large numbers of terms and natural language searching.

Mechanical to Holographic

This evolution recognizes the advancement from physical or mechanical objects to a virtual reality. For example, we are accustomed to visiting an actual library where we can sit down and examine a catalog, usually by sitting before a computer screen. We then can note the books or other items that we want and physically locate them. A virtual library enables us to browse the stacks on our computer screen, select the items, and check them out online. If they are physical items, they may be mailed to us; if electronic, we can download them.

A hologram is a lifelike, three-dimensional representation of reality. However, it is not real; it is virtual reality. The concept of a hologram is that a portion or part lets us see the whole; a piece contains all the elements of a whole. For example, a case study helps us to understand issues of a larger population, for example, studying a cohort to understand distance education issues.

We are seeing many examples of virtual reality in video games in which the player may be driving a race car or hunting wild animals in a jungle. With increased sophistication and diminishing cost of computer technology, additional simulations of reality may become an integral part of our lives as instructional or informational tools. For example, we might sit down at a tablet computer or cell phone to take a virtual tour of Disney World or the city of Seattle before we make a decision to take a family vacation. Tour books and brochures are giving way to virtual reality introductions to places we may want to visit, through electronic tour guides and virtual tours available online.

Determinate to Indeterminate

Determinate solutions are not possible in a complex society. Nothing has a definite answer. Facts that were once certain are no longer certain. The complexity in contemporary society leads to insoluble problems. There are few absolutes in today's world. For example, we ask questions about our society and seek answers that are not forthcoming. We see criminal behavior and want to know ways that such behavior can be curbed. We receive news reports of indisputable signs that our world is experiencing global warming. What are the causes? The United States engages in a military conflict in Afghanistan; how can the conflict be resolved so that troops can be returned home and Afghans can administer their own government? How can national debt be controlled? How can the size and cost of government be controlled?

The solutions of social problems may have been contemplated, addressed, and resolved more simply in past generations. For example, cures could be found for polio, smallpox, and other diseases. Public health issues were addressed through public policy that stopped the spread of disease. The American economy was more independent from the economies of other countries. Families were fairly stable, but now we have blended families in which father and mother bring together children from previous relationships. Even home life has become more complex.

As a result of the complexity in our society, the resolution of problems is much more complex. Few simple solutions are appropriate; solutions that were once determinate are now indeterminate. Solutions to social issues are likely to be short-term rather than long-term. The uncertainties of contemporary life can cause people to feel overwhelmed by the world and its problems.

Linearly Causal to Mutually Causal

The causes of phenomena have also become more complex. Mutually causal suggests that there is not a straight-line cause/effect relationship. Influence goes both ways, and a researcher can influence effects. Whereas $A + B = C$, we now may find that A plus B plus C interact simultaneously, dynamically, and nonlinearly.

To say that the home environment causes criminal behavior is inadequate. Criminal behavior may be caused by a congenital psychological defect exacerbated by home environment and social interactions with peers. Similarly, when an information professional is asked, "What is the population of Chicago?" we have an apparently simple question. However, does the client want the current population? How current? Does he or she want the city as defined by the city limits, or the metropolitan area? Even the simplest question is more complex than it appears.

While we may yearn for simplistic solutions to problems like poverty, crime, and the rising cost of living, such issues have multiple causes. However, politicians running for office still attempt to apply linearly causal solutions to mutually caused social problems. As information professionals, we must resist the urge to provide simplistic answers to complex questions and problems. We have a professional obligation to help our clients understand the complexity of issues by providing them with examples of various viewpoints.

Objective to Perspective

The term "objective" refers to things outside our thoughts and feelings. This term suggests the presentation of facts without bias. Although it is commonly suggested that we should be "objective" about an issue or a person, individuals are influenced by their experiences over time, and we form opinions about issues. Since we cannot be totally objective, let's recognize that we have a perspective, and it is an important perspective. Kuhn stresses that reality is a function of the belief system we hold. As information professionals, we must be aware of our perspective on issues and be willing to announce our biases when dealing with sensitive issues.

Summarizing the Paradigm Shift

The emerging view of reality is most succinctly described by the change in metaphor used to describe the image. The machine metaphor of the industrial age is being replaced by the hologram metaphor, which appears to be a truer likeness of the networked world. Information is distributed throughout the system; everything is interconnected, as manifested by the new concepts of outsourcing, insourcing, offshoring, open sourcing, supply chaining, and plug and playing, among others. These terms describe today's networked world, which is being reshaped and reconfigured every nanosecond and which, in turn, reshapes us. Herein lie the current societal revolution and the framework for understanding the impact of the Internet on information transfer and the information infrastructure.

The Environmental Context of Society

The information infrastructure is influenced by a variety of external factors. We call these factors "the environmental and social context"—in other words, the world around us. These variables influence us as individuals as well as all of the information transfer processes: creation, production, dissemination, diffusion, organization, and utilization of information. We have described these external factors in an earlier work (Greer, Grover, and Fowler 2013, 53–57).

The environmental and social context includes all of the following:

1. Culture—language, philosophical and moral values, educational system, concept of time, historical background, and all features that make up an organization's or community's culture

2. Physical geography—aspects such as climate and topographical characteristics

3. Political structure of society—the system for governance and underlying values regarding the role of government in a dynamic society

4. Legislation and regulations issued by legislative and regulatory agencies of government

5. The economic system under which the culture functions

6. Technology—the level of sophistication in terms of computer, tele-communication, and other technologies

7. Information policy—copyright laws, policies regarding secrecy, censorship, privacy, ownership, the public's right to know, and government responsibility to inform

A more detailed examination of these variables follows.

Culture

We often think of culture very broadly—the culture of an ethnic group or country. For example, the language, history, music, and dress of Brazil vary considerably from those attributes of Great Britain. Cultural differences also exist within countries. The local idioms, food, entertainment, history, and dress of rural Kansas vary considerably from the culture of Chicago or Los Angeles. The meanings associated with certain words vary, and attitudes also vary. These factors influence how people perceive information and how they think.

Physical Geography

The geography and climate of a community influence lifestyle, use of leisure time, and the culture of a community. California, Arizona, and the southern states boast a climate that encourages outdoor activities most of the year. While residents of Southern California, the Southwest, the Gulf Coast, and Florida are playing tennis or enjoying water sports on fall or winter weekends, those of us in the rest of the country are watching television, bowling, attending basketball games, or otherwise staying indoors. The weather influences recreational activities, and lifestyle differences influence attitudes.

Geographic features likewise inhibit or encourage travel and lifestyle. The congestion of cities is a stark contrast to the wide expanses of the American Midwest, where residents may drive 30 to 100 miles to the nearest city. Although technology overcomes distances, attaining broadband Internet service and the latest network technologies is still difficult for residents of remote areas in the United States and other countries in the world. Such difficulties have a significant impact on whether current information is available to these residents. Geography can influence both our perception and our access to information.

Political Structure

The governance system of an agency influences the flow of information. A hierarchy generally requires that information flows up the organization and down according to the rank of the individual. Those at the top of the hierarchy can successfully stop or change the flow of information as well as the content. An organization that is more informal usually encourages information flow in a casual manner without regard to status of the sender or receiver. In such organizations, information flow can be fast and unencumbered by status.

Political structure can also influence the content of information; for example, the values of a dominant political party may suppress information that is opposed to the prevailing views, and this suppressed information may find an outlet through "underground" or illicit channels. This political structure may be dictated by the culture of the organization and by the leadership style of those in leadership positions. Similarly, the format of the information may be dictated by the political structure; for example, print is preferred to electronic sources, or narrative is preferred to charts or diagrams.

Some leaders subscribe to an organizational philosophy that information must be controlled and that power or authority is derived from the control of information. Other leaders may believe that information should be shared among leaders and among all workers. These leaders will treat new information in a manner much different than those who wish to control it. In these examples, the philosophy of management focuses on the locus of control, and that control is determined to a large extent by access to information. How information is controlled and distributed is dependent on leadership style or philosophy. Political structure can be a major factor in the flow and content of information within and transmitted from an organization. Providing guidelines for the fair and equitable use of information in this age of technology is determined through laws and through information policies, as outlined below.

Legislation and Regulations

Legislation and regulations are issued by legislative and regulatory agencies of government. Where the political paradigm is controlled by the scope and limits of a constitution (e.g., the United States Constitution), freedom of access to information is at the core of this paradigm and is the bedrock of any information system designed for public use. The United States Constitution guarantees, through the First Amendment, the right of free expression of ideas, and other laws (e.g., copyright law) also protect individuals or groups who create new packages of information, so that others will not steal those ideas and call them their own.

Copyright protects various forms of expression, including songs and music, as well as prose and poetry. Similarly, patents protect the rights of inventors who create new products of various kinds, including hardware and software that may access, organize, store, or retrieve information.

A trademark is a recognizable design or expression that identifies services or products from a particular manufacturer or source. An individual, a business, or some other entity may own a trademark. Although trademarks are used in nearly every type of business, in the fashion industry a certain label or trademark can be very valuable in merchandizing clothing.

Still other laws, like securities laws, provide a framework for the exchange of stock information, determining when this information can be exchanged fairly. Securities laws also define what is considered illegal "insider" trading of information. Legislation and regulations are formal controls placed on the content and flow of information within an organization or a legal entity.

Information Policy

Information policy includes guidelines regarding secrecy, censorship, privacy, ownership, the public's right to know, and the government's responsibility to inform. Legislation is a formal policy adopted by a governmental body at the local, state, national, or international level and is addressed in the section above on legislation and regulation. Those laws and regulations that specifically address the creation, production, distribution, use, and storage of information may also be considered information policy. However, policy may be adopted within small and large agencies and groups of all types.

A policy is a generalization or general statement that provides guidelines for the transfer or use of information. For example, any social organization, school, library, college or university, business, or government agency at any level may articulate a policy for information use. Organizations may elect to keep no papers or financial records after seven years because the requirement for keeping tax records is typically seven years. Another organization, conscious of its historic contribution to a community, may elect to preserve all correspondence and donate it after five years to a local historical museum. While the practices outlined are much different, they are both policies that govern the transfer of information for their respective agencies and are consistent with their culture.

Economic System

An economic system supports a culture and social system financially. If capitalism is the basis for economic activity in a society, information is a commodity that appears in the marketplace as well as in the halls of the academies. Privacy and proprietary perspectives are very evident in this society. Other economic systems (e.g., communism) employ different approaches to the creation, reproduction, and distribution of information.

In a capitalistic society, the marketplace determines the marketability of a product. Since information is a commodity, the format and content of information is influenced substantially by the economic feasibility of the information package. For example, the marketability of a new information system, like the Internet, must be accompanied by affordable accessibility. The Internet was created in the 1960s by the U.S. government to exchange defense information, using state-of-the-art computers, software, and networking technologies of that time. This system was not available to the public until personal computers, telecommunications, and software became affordable to regional, state, and local government agencies and individuals. The widespread availability of the Internet became a reality in the mid-1990s with the confluence of technology and affordability.

The diffusion of the Internet into municipal governments, schools, colleges and universities, and individual homes took approximately 30 years. Often the diffusion process occurs first through the diffusion of information for recreation, followed by use for educational purposes. The use of film, audio recordings, compact discs, and digital video all found their way into the marketplace first as entertainment packages. To anticipate changes in technology applications, information professionals are advised to watch trends in the use of new technologies for entertainment as possible precursors to information use for other purposes. Cost, affordability, and packaging of information aligned with cultural values are key factors in the dissemination and diffusion of innovation and information.

Technology

The availability of computers, telecommunications, and other technologies becomes increasingly relevant to the processes of information transfer in a society. As noted above, the application of new technology to the transfer of information is dictated by economics. Affordability precedes adaptation on a wide scale.

Moore's Law helps us understand the affordability of technology. Gordon Moore, cofounder of Intel, made this prediction in 1965: "The number of transistors on a chip will approximately double every 24 months" (Intel). As a result, the scale gets smaller and smaller.

Technological innovation has transformed all aspects of information creation, recording, mass production, distribution, organization, storage, retrieval, and use. With rapid changes in technology, the transfer of information has been changed dramatically. The ability to transmit, store, and retrieve information on smart phones, tablet computers, and laptop computers, along with the availability of instant messaging and social media, has had a major impact on daily life and on business and professional practice.

Technology has had a significant impact on education at all levels. Distance is overcome in public schools and in higher education through videoconferencing, enabling learners in remote areas to participate in learning experiences previously unavailable to them.

Because the Internet is ubiquitous, control of information is no longer possible. Control is a desirable attribute in the Industrial Age, but control is not possible with the access to information dispersed through the Internet. Although the governments of China and other countries have attempted control of information, smart phones and social media have enabled people to break through the controls that were imposed. The above variables influence each other and, in turn, influence various elements of the information infrastructure.

Culture Shapes Values and Paradigms

Previously we defined "culture" as all of the elements that make a society unique: language, history, music, dress, and customs that have evolved over time. These elements of culture are imbued with values that influence how people perceive information and how they think. Consequently, the worldview that evolves from a culture is a paradigm, which shapes how people in that culture perceive their world, their communication, and the information that they receive from people within and outside the culture. The values that are accepted within the culture serve as filters for the information that individuals and groups within the culture receive.

The environmental context previously outlined is those factors that interact within a culture to modify the culture over time—a continuous, shifting process of change. And culture is not monolithic—the people in a culture have differences in their values and preferences; the specifics of their culture may be influenced by their environmental context, which in turn influences an individual's perception of the world and acceptance or rejection of new information and knowledge. The process is ongoing and complex. Within a culture, several paradigms may exist, and each may evolve; as paradigms change, they in turn influence culture and the information infrastructure. These issues will be explored further as we delve into the various components of the information infrastructure.

Summary

We have examined the historical roots of the technologies that have led to the cataclysmic changes that have occurred in our society. We examined models that explain these fundamental changes and described the environmental factors that influence how information and knowledge are created, disseminated, diffused, and utilized. The models introduced are signposts to guide us through the journey of discovery as we travel through a world dominated by digital content.

References

Achleitner, Herbert K. and Martha L. Hale. 1988. "Information Transfer: Educating Information Professionals in the Emergent Paradigm." In *Proceedings of the First Joint Meeting Between the Association Internationale des Écoles des Sciences de l'Information and the Association for Library and Information Science Education*, ed. Rejean Savad. Montreal, Canada.

Capra, Fritjof. 1996. *The Web of Life: A New Scientific Understanding of Living Systems*. New York: Anchor Books.

Intel. Moore's Law and Intel Innovation. http://www.intel.com/about/companyinfo/museum/exhibits/moore.htm.

Kuhn, Thomas S. 1970. *The Structure of Scientific Revolutions*. Chicago: University of Chicago Press.

Schwartz, Peter and James A. Ogilvy. 1979. *The Emergent Paradigm: Changing Patterns of Thought and Belief*. Menlo Park, CA: SRI International.

3

How Paradigms Influence Society in the Information Age

Chapter Overview

The information infrastructure is complex and has evolved over time. This chapter explores the components of the information infrastructure, including the transportation, communication, and technological components. The library and information profession is examined as a vital contributor to this infrastructure. The evolving new paradigm is examined as a framework for analyzing the information infrastructure and for understanding the current, sometimes confusing changes occurring in the creation, dissemination, and utilization of information in contemporary society.

In Chapter 2 we introduced characteristics of the emergent paradigm created by Peter Schwartz and James Ogilvy (1979), applying it as a lens for understanding the digital world. These characteristics describe the collapse of certainty in society evident in the intellectual revolutions taking place in the sciences, humanities, and social sciences. In summary, contemporary society is described as complex, heterarchical, holographic, indeterminate, mutually causal, morphogenetic, perspectival, and digital.

Table 3.1 is a matrix that applies the Schwartz and Ogilvy characteristics as a way of describing the evolution of the information infrastructure from its old-paradigm values and systems to the emergent values and systems of the digital age. The matrix provides an overview of the elements of the information infrastructure as it has evolved. The left column represents the old paradigm, while the right column represents the information infrastructure in the contemporary digital age. A brief history of this evolution follows.

Table 3.1   How the Emergent Paradigm has Changed Library and Information Services

Old Paradigm	Emergent Paradigm
Simple Materials and processes were simple, manual, and understandable.	**Complex** Advancements in technology have produced a variety of methods for recording, mass-producing, and disseminating information.
Hierarchy The organizations in the processes were clearly organized with a hierarchical structure.	**Heterarchy** Lines of authority are unclear. Technology has enabled individuals to circumvent hierarchical structures, providing much freedom of expression.
Mechanical Processes for recording, mass-production, dissemination, and use of information used manual or simple machinery.	**Holographic** With the advancement of technology and application of digitization, visual information could be three-dimensional, opening new avenues for expressing ideas.
Determinate Solutions for problems could be concrete, explained, and certain. Facts were known and remained stable.	**Indeterminate** The world is complex, and solutions are likewise complex; furthermore, many problems have no definite solutions, and "facts" change as new discoveries reveal new insights.
Linearly causal Just as solutions to problems could be explained, causes and effects also were direct and easily seen.	**Mutually causal** Complex problems or issues usually have many causes that interact. Cause and effect are not easily explained.
Objective Objectivity suggests the observation of information without bias.	**Perspective** In the new paradigm, recognition is given to the notion that humans harbor a unique perspective on their perceptions of information. Objectivity is not possible.
Analog Processes for creating, mass-producing, disseminating information were mechanical and used relatively simple technologies. Information was printed manually or mechanically on paper or other material.	**Digital** Processes are digital, applying computer-aided technologies in all of the information transfer processes. These technologies enable the transformation of the information transfer processes.

Evolution of the Information Infrastructure

The information infrastructure includes many components, as noted in our definition in Chapter 1:

> The information infrastructure is a **global** network of **people, organizations, agencies, policies, legislation, processes, and technologies** organized in a loosely coordinated system to enhance the creation, production, dissemination, organization, diffusion, storage, retrieval, and preservation of information and knowledge for people. The primary objective of this network

is the diffusion of knowledge for a society. (Greer, Grover and Fowler 2013, 100)

As a first step in exploring this infrastructure, we will trace the evolution of several components of that infrastructure.

An understanding of the information infrastructure can be enhanced by reviewing the evolution of our systems of creating, recording, and transmitting recorded communication and the history of libraries and information services. The evolution of the various components of the information infrastructure was driven by the external environment: such factors as culture, economics, policies, government regulations, and technology. These factors drove improvements in transportation as well as improved methods for recording, reproducing, and distributing information.

As information sources increased in volume, the need arose for personnel to organize and retrieve this information in organizations that became known as libraries. As libraries evolved to meet the changing needs of society, the need for specially trained professionals also evolved. Both libraries and library and information professionals have been significant components of the information infrastructure. This section explores that evolution of the transportation infrastructure before the digital age.

Transportation

A fundamental support system for the transfer of information has been the transportation infrastructure. We present an overview of transportation history to show that support for information transfer.

The usual definition of "transportation" is the movement of goods and people from one destination to another; however, information often is transferred as well, often orally. Throughout human history, transportation infrastructure has played a vital role in the shaping of a culture. Grant (2003, 1) stated, "Every place has a diverse transportation story, replete with failures and triumphs. Individuals who explore the transportation heritage of their community will quickly discover that various forms of transport have had a vital impact on ordinary life."

Transportation provides linkages for economic and social growth. Without transportation, a community suffers from isolation. The presence of a railroad in a town in 19th-century America brought workers and economic stimulus, including boarding houses and services for railroad workers and passengers. Similarly, canals, highways, and airports have had the same economic and social impact on communities. When superhighways were constructed beginning in the 1950s, communities that were bypassed often struggled to grow or even survive.

Use of Animals

Human use of machines for transportation has occurred only during the last two centuries. Before horses, the railroad, and automobiles, humans delivered information person to person; they carried goods on their heads or backs. Eventually, animals were trained to draw vehicles or carry goods on their backs. According to Dent (1972), evidence exists that the first animal to be domesticated and used for transportation was the dog in the Middle Stone Age.

Domestication of large animals probably began in about 7000 BCE in order to provide a food supply. For much of the history of roads, animals provided the means of power for travel. The best-known animals used for transportation have been cattle, donkeys, dogs, goats, horses, mules, camels, elephants, buffaloes, llamas, reindeer, and yaks; humans also have been employed in transportation.

The horse was originally a pack animal but became a riding animal and was harnessed to a war chariot used by the aristocracy in Western Asia, Europe, and North Africa from 2000 BCE to about 500 CE (Dent, 6–7). A horse and rider could travel about 60 to 80 kilometers per day and carry information. The Persians in the reign of Cyrus the Great in 550 BCE established post houses and stables along a route so that tired horses could be exchanged. Teams of horses and riders could travel as much as 250 kilometers per day. The Romans and Chinese also used this method for sending messages. A similar messenger system of relays was used at various times, and a master of posts was appointed in England in 1509.

In 1860 this relay system was employed by the Pony Express, which carried mail from Missouri to California, the riders traveling about 300 kilometers per day. Riders changed horses every 10 to 30 kilometers. The Pony Express operated for only 18 months and was made irrelevant by the transcontinental telegraph.

Roads

Roads have been a vital part of the transportation infrastructure that supported the information infrastructure. Lay (1992, 5) asserts that the first roadways or pathways across the countryside were created by animals migrating to seek water and salt. Humans in temperate zones were using pathways by about 10,000 BCE. A growing civilization increased social and economic pressures to construct pathways, especially through difficult terrain. These trails usually led to campsites, food, and water.

Roads developed from pathways were haphazard and bore little resemblance to roads as we know them. As countries developed, road systems were part of the government structure. The elaborate road system of the Roman Empire is a good example of early road development to facilitate travel and communication of information.

In North America, trade routes initiated roads that opened the West for settlement. An example is the Santa Fe Trail that linked the Missouri towns of St. Joseph and Independence to Santa Fe, New Mexico. This trail and the Oregon Trail were used extensively until the establishment of railroad routes in the second half of the 19th century. As the nation moved west, so did the need for communication of information.

The Land Ordinance Act of 1785 divided land outside established cities into 36-mile squares, which were divided into 36 square-mile units. The squares were quartered into farms, and each farm was required to allocate a 33-foot strip of land to create a 66-foot roadway that followed the property boundaries at a one-mile spacing. Hence, plans were laid for a national road system. Following World War II, the U.S. government established a long-term plan for implementing an interstate road system that speeded the transport of people, goods, and mail to all parts of the country.

Waterways

From ancient times, rivers, lakes, and seas have served to transport goods and passengers. These waterways were often augmented with the construction of canals. In the United States, the canal served as a transition between animal-powered vehicles and steam-driven railroad cars. During the early 19th century, about 3,700 miles of canal were constructed, mostly in New York, Pennsylvania, and Ohio (Grant, 51). These canals served as an inexpensive, reliable means of transporting goods, augmenting navigable rivers and lakes. As railways were built and better locomotives were built, the canal was superseded as an efficient mode of transportation, although some canals are still used regularly in the twenty-first century.

Ships

Humans' venture onto water is obscured in history. Evidence found in illustrations in tombs and on decorated pots suggest that where there was timber, floating logs may have been the inspiration for shaping logs, hollowing them out, and forming primitive boats. When timber was not available, papyrus reeds were bound together to build watercraft; the Egyptians are known to have built sailing ships as early as 6000 BCE (Macintyre 1972).

The Phoenicians, and later the Greeks, engaged in trade using ships to import goods from distant ports in the known world. As the cradle of European civilization, the Mediterranean has more than two thousand years of shipbuilding history. These ships undoubtedly carried information as part of their cargo.

Sailing ships provided transportation from the time of the ancient Egyptians until the 18th century, when numerous experiments were conducted to employ a steam engine to power a paddleboat. The first successful steam-powered boat that was feasible for shipping on rivers or canals was designed by American inventor Robert Fulton. Fulton's boat, the *Clermont*, was launched on the Hudson River in August 1807. The steam-powered craft traveled up the Hudson River 110 miles from New York City, proving its durability. Fulton built several other steam-driven boats to carry goods on the Hudson River and extended his business to the Mississippi River as well. His ship the *New Orleans* traveled from Pittsburgh down the Ohio and Mississippi rivers to New Orleans in 14 days. Steam-powered boats soon were launched on other rivers in the East and Midwest.

The engines and ships were re-engineered for ocean travel, although the first ships used the engines to augment their sails. The Queen Steam Navigation Company was the first to cross the Atlantic Ocean with only a steam-powered ship in 1838, a journey of 17 days (Macintyre 1972).

Ships continued to employ new technologies to produce larger and more efficient engines that enabled the transport of goods, passengers, and mail long distances over bodies of water in shorter amounts of time. Ships were the major vehicles for long-distance travel until superseded by air travel in the 1930s.

Railroads

Railroads provided a breakthrough in transportation and can be traced to the mines of Austria and Hungary of the 15th century. This primitive system used two planks laid side by side with a gap between and secured to the mine floor. Iron rails were first used in England in 1767. Cast-iron plates were laid on top of the wooden rails that already existed. Horses provided the power for the cars, and iron rails increased tremendously the horse's hauling capacity (Georgano 1972).

By 1800 the concept of railways was well known, and by this time it was becoming apparent that the hauling capacity of railways, even when horses were the haulers, rivaled or surpassed the speed and economy of shipping via canals. When mechanical power replaced the horse, railways began to grow in use.

The first main-line railway in the United States was a section of the Baltimore and Ohio in 1829 using horses to pull the cars. By 1831 this railroad began experimenting with the use of steam engines.

The commercial success of these early railroads generated interest in railroads worldwide. Governments became involved in planning and regulating rail routes. In the United States, railroads played a fundamental role in the opening of the West, transporting goods, passengers, and mail. With improved roads and with the advent of air travel, railroads have remained an efficient and economic system for shipping products long distances over land.

Cars and Trucks

By the end of the 19th century, use of the horse for transport was viewed as problematic:

> Commentators have concluded that, at the time, the developed world had reached horse capacity and any appreciably greater numbers of horses would have been quite literally insupportable, and that the motor vehicle had raised the quality of urban life by driving the smell and squelch of the horse from the streets. (Lay 1992, 167)

Early engines were used to transport products, but Richard Dudgeon of New York built a steam-powered car in 1858. Although forty or more steam cars were built between 1860 and 1880 in the British Isles, none was mass-produced.

In 1860 Étienne Lenoir of Paris patented an internal combustion engine that burned gasoline. Georgano (1972, 29) credits Karl Benz as the father of the automobile in 1885. "Benz planned his car as an organic unit of chassis and engine, not as a conversion of a horse-drawn vehicle, but, more important, it had direct descendants which were on sale within a few years" (Georgano 1972, 29).

Production of motorcars increased from two in 1886 to eleven thousand by 1900 (Lay, 168). During the 1906 San Francisco earthquake, cars were used to provide emergency services, and their use was given a substantial amount of publicity. Cars and trucks made a significant contribution during World War I, and car usage increased after the war.

By 1904 Detroit was producing 20 percent of cars made in the United States. Ransom Olds began mass production of the Oldsmobile in 1902, and Henry Leland produced a Cadillac with completely interchangeable parts. Henry Ford introduced the Model T in 1908 and began mass production of the car. By June 1909, 100 cars per day were being made. Ford introduced the moving assembly line in 1913, and Ford sold 300,000 cars in 1914. Yearly production of Model Ts surpassed one million in 1922.

Early trucks also were steam-powered; Gottlieb Daimler built the first truck powered by an internal combustion engine in 1891 and began manufacturing them in 1894. Although horses were still in use for transport, by 1910 the truck was considered superior to horse-drawn wagons where paved roads and streets were available.

The French first used trucks in the military in 1897. Trucks were used in World War I by all sides, and the United States used one- and three-ton trucks that were called "Liberty Trucks." After the war the U.S. government sold many of these vehicles to citizens, providing an impetus for the conversion to motorized vehicles.

Buses powered by internal combustion engines were first produced by Karl Benz and operated in Germany in 1895. By 1910 motor-driven buses had replaced most horse-drawn buses. The last horse-drawn buses stopped running in London in 1914 (Lay 1992, 173).

Conversion to gas-powered cars and trucks was not without opposition. People complained of the noise and smell produced by the engines and the dust produced by the vehicles on unpaved roads. By 1905 the United States built the same number of cars as carriages. In 1909 the car replaced the horse for the transport of U.S. presidents. With the paving of roads and streets, as well as advancements in the gas engine and inexpensive production techniques, the automobile grew in popularity and use during the first three decades of the 20th century.

Cars and trucks enabled transportation of people, goods, and mail over great distances and to remote parts of a nation. As people and mail traveled, so did ideas and information.

Aviation

On December 17, 1903, Orville and Wilbur Wright made their historic flight at Kitty Hawk, North Carolina. By mid-century, air travel had become a major form of commercial transportation. Aircraft development occurred in France, England, and Germany. Frenchman Louis Blériot was the first to cross the English Channel by air in 1909.

During World War I, improvements were made in airplane structure and in engine design. In 1920 the Aeromarine Sightseeing and Navigation Company flew passengers between New York City and resorts as far away as Florida, Havana, and Nassau (Grant, 153). In 1918 U.S. Army pilots began flying planes to deliver mail, and in 1925 Congress passed and Calvin Coolidge signed the Contract Air Mail Act, which authorized the postmaster general to select air routes for delivery of mail. In Europe, too, air travel was growing in popularity and mail was being delivered by airplane. Charles Lindbergh's successful flight from New York to Paris in 1927 helped to spur long-distance air travel.

By 1930 air travel was becoming more commonplace, and in May 1930, Boeing Air Transport employed the first airline stewardesses on a flight from San Francisco to Chicago. A coast-to-coast flight was then possible in about twenty hours.

The jet engine was developed in Great Britain in 1937, and a prototype jet plane was first flown in August 1939. The first jet airline service was initiated in May 1952 on the London to Johannesburg route by BOAC. On October 26, 1953, Pan American Airways made the first American transatlantic jet-powered commercial flight in a Boeing 707. With the jet-propelled aircraft, transcontinental travel was reduced to approximately five hours. Now airlines enable the delivery of mail and products overnight from distant cities; aviation has become an important element in the information infrastructure.

Postal Services

Historical records indicate that postal services were developed as early as 2000 BCE in Egypt and during the Zhou dynasty of China in about 1000

BCE. A vast postal system, the *cursus publicus*, was developed during the Roman Empire with its excellent roads and relay system (*Encyclopedia Britannica*).

The postal system in colonial America progressed through the leadership of Benjamin Franklin, who was Philadelphia postmaster and became the postmaster for the American colonies in 1753. Franklin created an extensive and speedy (for that time) mail service within the Colonies and England. He later was appointed postmaster general for the United States in 1775 and was responsible for building a foundation for the postal service in this country, a vital infrastructure for information dissemination.

The postal service was a major user of all transportation systems as they developed. Trucks and airlines eventually superseded the stagecoach, Pony Express, steamboat, canals, and railroads. The sorting of mail in transit was introduced in 1862, and railroad mail service became the dominant form of mail delivery into the 20th century.

Postal services expanded during the 19th century to include registered mail (1855), money order service (1864), special delivery (1885), parcel post and insurance services (1913), and certified mail (1955). Mail was divided into three classes in 1863, and a fourth was added in 1879. Although mail delivery has been expedited with airmail, in recent years the postal service has experienced a decline in use because of the telephone and, most recently, electronic mail and texting.

How the Printing Press Revolutionized the Information Infrastructure

The increase of literacy among the growing middle class led to an increased demand for books; however, hand copying was time-consuming and could not accommodate the demand. Johannes Gutenberg was able to combine advancements in paper presses, metalwork, and engineering to develop a printing press with movable type. Gutenberg improved the printing process by refining both typesetting and the printing process.

Another important development that facilitated transfer of information was evolution of the codex format, which originated during the Roman Empire. The codex, the book form we are familiar with, proved far superior to scrolls because pages could be easily turned and it was more compact and less costly.

Other economic factors were the capability of printing on both sides of a page in the codex format and the development of thinner paper to replace parchment. The availability of paper and an improved process of printing revolutionized the reproduction, mass production, and dissemination of information at a time when a growing middle class needed education and materials to enable widespread teaching and learning.

Mass Production and Spread of Printed Books

The invention of mechanical movable-type printing led to a large increase in printing activities across Europe. Printing spread to cities throughout Europe by the end of the 15th century. European printing presses in 1600 were capable of producing 3,600 impressions per day, compared to movable-type printing in Asia, where printing was done by manually rubbing the back of the paper to the typeset page, which would produce forty pages per day. The improvement of publishing potential resulted in faster dissemination of ideas. More than 750,000 copies of Erasmus's work were distributed during his lifetime. Martin Luther's works were distributed in 300,000 printed copies. The improvement in speed and quality of the printing process led to the publishing of newspapers for transmitting current information to the public.

Newspapers

For more than two centuries, the newspaper has served as a staple in the information infrastructure, and it has been convulsing with the immense change brought by new technologies and communication media. A newspaper is a periodical publication that usually contains news of current events, informative articles, feature stories, editorials, and advertising. Traditionally it has been printed on low-quality paper called "newsprint" but now is distributed also (or exclusively) in digital format via the Internet.

Early precursors of newspapers can be traced to ancient Rome's government announcements carved in metal or stone and posted in public places. In China, news sheets produced by the government circulated among court officials during the late Han dynasty of the second and third centuries CE. During the eighth-century Tang dynasty, a formal handwritten publication that disseminated government news was published and circulated to intellectuals and government officials. By the Ming dynasty (1367–1644), news publications circulated to a wide circle of society. "Even though the Chinese had produced the essential technical prerequisites of the newspaper in its European guise before the year 1500, the Chinese press was very slow to develop" (Smith 1979, 14).

The development of the printing press in the late 15th century spawned in Europe printed accounts of single news events. Smith (1979) identifies the publication of news leaflets in Vienna, Poland, and Paris. By the end of the 16th century, numerous printers were publishing news leaflets and pamphlets on an irregular basis.

Johann Carolus is usually credited with publishing the first newspaper in Strasbourg, Germany, beginning in 1609. Smith (1979) credits Holland and Germany as the countries that gave root to the newspaper. Towns along trade routes were first to establish newspapers: Cologne in 1610, Frankfurt am Main in 1615, Berlin in 1617, Hamburg in 1618, and several other towns in the 1620s.

As the 18th century began, improvements in the British postal system made daily publication and distribution practical. The first such daily newspaper was the single-sheet *Daily Courant* (1702–1735).

In Boston in 1690, Benjamin Harris published the first newspaper in the American colonies, *Publick Occurrences*, but the British government suppressed it after one edition was published. *The Boston News-Letter*, initiated in 1704 by John Campbell, was the first continuously published newspaper in the colonies. Soon after, weekly papers were published in New York and Philadelphia. *The Pennsylvania Evening Post* was the first American daily newspaper, first published in 1783.

New developments in technology influenced printing techniques and newspaper distribution. In 1814 the steam-driven "double-press" was introduced at the London *Times*, increasing production to 5,000 copies per hour. Consequently, the *Times*'s circulation rose from 5,000 to 50,000 by the 1850s. The invention of mechanical lead type, curved printing plates, automatic ink-feeds, and the cylindrical rotary press all contributed to faster publication and distribution of newspapers.

A major breakthrough was the introduction of automatic typesetting with the linotype machine, which was operated by a keyboard to automatically set and *justify,* or fill a line of text by adjusting the space between words. The first linotype machines were used at the *New York Tribune* in 1886. Electricity was introduced in 1884, as were machines that cut, folded, and bound newspapers.

By 1900, the newspaper was established worldwide as an essential link in the information infrastructure. Now, in the 21st century, the newspaper reflects the emergent paradigm by appearing in paper and electronic formats, with opportunities for readers to comment on articles. Newspapers have become a two-way communication channel for the dissemination of information.

Communication Infrastructure

Since information transfer by definition is recorded information, it is helpful to explore the history of recorded messages. Meadows (2006, 1) provides a standard definition of communication: "Communication is based on the transmission of symbols and the interpretation of these symbols by the receiver of the communication." Symbols may be sounds, including spoken words, pictures, digital symbols (letters, numbers), odors, tastes, and feeling. In other words, humans communicate using all of their senses.

Primitive Messages

Recorded messages (drawings) thought to be 50,000 years old have been discovered in caves in France and Spain. These drawings were iconic—pictures of objects or animals—and it wasn't until about 3300 BCE that the Sumerians recorded symbols on the flat surfaces of clay tablets. Pictorial symbols gradually evolved into abstract symbols that did not look like the subjects they represented. The English alphabet was created by Semitic peoples who lived in the region of present-day Lebanon, Israel, Palestine, and Egypt.

New media for recording the symbols evolved: papyrus and animal skins; parchment was made from sheepskin and vellum from calfskin. Paper was invented in China in about 100 CE Ink was used as early as 2500 BCE.

During the Middle Ages (500–1500 CE), classical European learning was supported by the Christian church. The Renaissance in the 14th century and the Reformation of the 16th century resulted in more people reading and writing. These efforts to read were supported by the invention of the printing press in about 1450 by Johannes Gutenberg. The printing press became an important part of the information infrastructure, as it enabled the fast reproduction of newspapers, magazines, and such monographs as books and pamphlets.

From the time of Caesar in the Roman Empire to the beginning of Queen Victoria's reign, all major improvements in long-range communications were in transportation (Meadows, 29). The printing press expedited the production of books, but they had to be transported physically to reach a larger audience. The invention of railroads shortly after the start of the 19th century was an important impetus for the conveyance of recorded information. Before railways, ships were major transporters of information. Distribution of printed publications utilized various forms of transportation. Following the development of writing and written communication came mail services.

Photography

Photography, recording visual images on a surface such as a sheet of paper, was popularized with Frenchman Louis Daguerre's invention of the daguerreotype in 1829. This method used copper plates coated with silver iodide, but ten years later in England, Frederick Archer used glass plates covered with different chemicals; this proved to be a much more practical process. In 1889 Thomas Edison invented a way of projecting photographs of a moving subject, resulting in the illusion that the images were moving—the birth of movies. In 1888 George Eastman began using paper coated with emulsion to make photographs, replacing the bulky glass plates previously used, and Eastman Kodak Co. was formed. Photography soon became a means of recording family events and news events, and along with motion pictures, it became the best way of communicating visual information until invention of television.

Telegraph

Electricity enabled the invention of the telegraph, a system of sending coded messages instantaneously over wire. While the Pony Express carried a message by horseback from St. Joseph, Missouri, to Sacramento, California, in a week, the telegraph could send the message in a matter of seconds. Although invented in several places, good working systems were used in England in 1837 by the Great Western Railway and in 1844 in the United States by Samuel F. B. Morse. The many telegraph companies initially formed soon consolidated into the Western Union Telegraph Company, which is still in use. Except for sending money, most other functions have been assumed by fax, electronic mail, and various forms of social media.

Messages sent by telegraph had to be coded and transmitted from an office and transmitted via lines to a distant office. Telegraphs could not, of course, carry the human voice or music. Alexander Graham Bell developed a machine that could carry the human voice over wires and patented it in 1876.

Telephone

The telephone provided many features that the telegraph did not. The human voice could be transmitted without any coding, and anyone could use the phone. Although phones needed to be linked by wire, telephones could be linked to a central switching station; soon it was possible to link stations, and the telephone infrastructure was created, and people in distant cities could be linked.

Telephone use grew quickly, and Bell's Telephone Company was eventually superseded by the American Telephone and Telegraph Company (AT&T) in 1885. AT&T has done quite well since, updating technologies as they became available. Now AT&T and other companies provide one-stop "bundled" services for cell phones, landlines, and Internet connection.

Radio

Michael Faraday, James Clerk Maxwell, and Heinrich Hertz were scientists whose experiments with electromagnetic waves led to Guglielmo Marconi's invention of a radio that could transmit across the English Channel. In 1901 Marconi conducted an experiment that transmitted a signal from Cornwall, England, to Newfoundland, a distance of 2217 miles. By 1904 radio was installed in 124 ships at sea. Although radio waves were used, the signals in these early days were sent in code; it was wireless telegraphy, with the same limitations as the telegraph, requiring senders and receivers who knew the code.

David Sarnoff, a Marconi engineer, saw the commercial possibilities for radio transmitting voice and music. In 1918 Marconi's company and General Electric agreed to a merger, forming Radio Corporation of America (RCA), with Sarnoff as the head of the company. The first commercial broadcast was by Westinghouse Electric Company in 1920, carrying news of the U.S. presidential election. Within a decade, radios and radio programming spread throughout the United States and became a communication vehicle for news and entertainment unmatched until the invention of television.

Television

Concurrently with the development of radio, the technology for television was evolving. In 1929 the British Broadcasting Corporation began broadcasting, and in 1939 the National Broadcasting Company, part of RCA, transmitted the first television transmission, reporting the opening of the 1939 World's Fair in New York City. Development of television was postponed until the end of World War II. By that time, the economy had improved after the Great Depression, and families had money to buy televisions. Although color television had been produced as early as 1928, the technology was not developed sufficiently until the U.S. government approved it in 1953.

Government regulation was required for television transmission because the frequency spectrum is in the public domain, and frequencies must be controlled so that only one transmitter uses a specific frequency, which government assigns. Also, broadcasters are regulated so that they do not change their method of transmission in ways that would render receivers useless.

Television signals originally were broadcast over the air by antennae placed on tall buildings or mountains and then transmitted by cable to homes and offices in the area—community antenna television (CATV). More recently, the signals have been transmitted via satellite to community antennae and delivered via cable. In recent years, television programs have been transmitted via the Internet to computers and cell phones.

The Internet

Computers alone cannot transmit information; however, computers have enabled the establishment of the worldwide communication network we know as the Internet. The U.S. Department of Defense initially sponsored computer science research for a communication system that linked computers. The Advanced Research Projects Agency (ARPS) conducted the research that developed a network of computers called ARPANET in 1969. The network linked computers via telephone lines at research facilities in the United States and enabled the transfer of files among computers—communication that became known as e-mail.

ARPANET was expanded beyond government agencies to private companies, and in 1983 an international set of standards (Internet) was implemented, and that term became used to describe the worldwide network. Key components are the Internet service providers (ISPs), which provide individuals access to any other computer on the network. While some people may call the Internet "the global information infrastructure," we submit that it is only one component (albeit a very important one) in the information infrastructure.

Communication Satellites

A relatively recent addition to the global information infrastructure is the communication satellite. First proposed in a scientific paper written in 1945, the earliest satellite was launched by the U.S. Air Force in 1958. To be effective, a satellite must be in geosynchronous orbit, which means it moves around the earth at the same rate that the earth turns, so that the satellite appears to be stationary. Meadow (2006, 84) describes a communication satellite as being "like a very tall tower that can receive radio signals and retransmit them." Most satellites are about 22,300 miles above the earth.

Wireless Telephone

The wireless telephone or cell phone was invented in 1947 as a mobile radio connected to a telephone system. Range for use was very limited, but the first cellular phone system was initiated in 1979 in Tokyo. The U.S. government approved the use of cellular phones in the United States in 1982. Cell phones provided the capability to connect with the Internet, send and receive text messages, download software, receive and send pictures, and play recorded music, as well as providing telephone service.

Currently the cell phone has evolved into a multipurpose information receiver, recorder, and disseminator. Most smart phones have the capability of taking still photographs and video motion pictures. They can send and receive text messages as well as telephone calls. Cell phones can access the Internet as well as send and receive e-mail. It is now possible for people to communicate with others wherever they are able to receive radio signals and to receive and transmit oral messages, photographs, and motion pictures. As noted above, television programming can also be received via the Internet on cell phones.

Information Infrastructure Evolution Summary

The information infrastructure includes a wide array of technologies, personnel, and conveyances. The transfer of documents and printed resources requires efficient transportation systems. The transfer of digital information requires sophisticated technologies to transmit information instantaneously worldwide. Information transfer dates from the primitive drawings by prehistoric humans and was enhanced immeasurably by the invention of the printing press in the 15th century. Since then, the improvement in transportation and communication technologies has resulted in a proliferation of media.

The merging of communication technologies has provided easier access to information consumers. For example, through the capabilities provided by communication satellites and wireless telephone technologies, people throughout the world have access to the Internet via their cell phones. Smart phones combine the telephone with computers, photography, radio, television, and audio/video recordings. Furthermore, through these communication technologies, individuals may participate in the creation and dissemination of information; the process is no longer one-way communication to the consumer. We are seeing the emergence of a new paradigm in information transfer. The complexities that accompany this emergent paradigm in information transfer require a knowledgeable profession to organize and disseminate it, the province of the library and information professions.

Libraries Evolve within the Information Infrastructure

In this complex world of competing and converging technologies, librarians play a vital role as mediators and leaders in this swirling array of fast technological change. The major role of the librarian was to catalog or arrange items for retrieval. We can generalize that a civilization progressed when it developed a written language and archived records from its past. Primitive societies lacking an alphabet were unable to create a record of events as they developed. The consequence was a lack of access to the past beyond the human memory for their myths, songs, and folk customs. Without archives of their past, they could not advance beyond the knowledge that could be remembered, and they tended not to advance to a sophisticated state.

Historical Development of Libraries

Libraries were not widespread and available to the masses until the Industrial Revolution, which brought the rise of cities, industry, and printing. We can trace the history of libraries back 5,000 years and categorize ancient libraries into four groups: (1) government, (2) religious, (3) commercial, and (4) private or family libraries.

The first, government libraries, collected the treaties, legislation, and genealogy of the royal family, lines of succession, and other recorded documents pertinent to governance in a variety of formats; for example, clay tablets, papyrus scrolls, and parchment. The information was recorded manually, and the information sources were arranged physically by subjects, often in different rooms.

Religious libraries contained rules for the education of the clergy. The creed of the religion was recorded and stored—the beliefs, sacred writings, and rituals of the religion or sect.

Private libraries archived the private papers of families, including records of property holdings. They also held records of the relationships between the family and rulers or government of the day.

Commercial libraries stored and organized the records of commerce. This included business transactions, business papers, and other records of various kinds.

These same four library categories continued to reflect the archival function of libraries throughout history. For example, the medieval library model was the monastery library, a treasure intended to be valued equally for the collection of rare manuscripts that were copied carefully and traded, and for the enlightenment that its collection provided.

The royalty and the privileged classes reserved early libraries for their use. The Romans created "public libraries," but not as we know them today. They were available for use by the wealthy classes, but they were maintained publicly instead of by individuals. At the time of Julius Caesar's death, a number of public libraries were open to the upper echelon of Roman society. Libraries have been a key component of the information infrastructure throughout the history of civilizations that recorded their thoughts and organized the materials on which these thoughts were recorded.

With the availability of printed resources, libraries were established throughout Europe and in the colonies of North America. During the 18th century, national libraries were established; for example, the British Museum was founded in 1753 and the Bibliothèque Nationale in 1368, tracing its origin to the royal library of Charles V. This collection became the National Library of France in 1792.

In the United States, Harvard University and its library were established in 1636. Yale University was established shortly after (1640); William and Mary College followed in 1693. The creation of the first PhD program at Harvard in the 1860s required library collections to support research. Another significant change was the introduction of majors and elective courses in college curricula, allowing faculty to teach topics related to their interests. This innovation placed a budget strain on academic libraries wanting to support an expanding university curriculum. In response, the library became the laboratory for the social sciences and the humanities.

In 1731 Benjamin Franklin established the Library Company of Philadelphia, the first public subscription library. It exists today as an independent, nonprofit research library. In the eighteenth and nineteenth centuries, mechanics institutes and mercantile libraries were founded to provide opportunities for workers to educate themselves and occupy management positions in expanded industries. Many of these institutions remain in service today.

In the 1830s, New York Governor DeWitt Clinton promoted the passage of enabling legislation to create school district libraries. This legislation was later amended to allow public funds for the development of these libraries. These publicly funded libraries were intended as public libraries, the first instance of government money used for libraries. In the 1850s public libraries were founded in Boston, New York, and Philadelphia. The Boston Public Library was the first such library established for the common person.

As with the mechanic and mercantile libraries, a major motivator for the development of the public library was to support formal education and self-learning. While community leaders wanted to support self-education, libraries also sought opportunities to acquire the best books, the cultural function of libraries. Although collecting for the purpose of educating clientele and enhancing the culture of patrons, librarians continued to place a high value on the collection, and the values of the archival function were maintained; that is, books were treasures that must be preserved.

Library and information service eventually evolved to address the needs of clientele in response to society's needs. This paradigm shift from valuing the collection to valuing service to clientele is discussed later in this chapter.

The Library Profession

Until the latter part of the 19th century, people managing libraries were part of the intelligentsia: the clergy, professors, historians, and other educated citizens. As with Benjamin Franklin, these were people who respected books and reading. A good part of the development of libraries was due to the work of printers and publishers, whose interest was to encourage book use. The combination of scholar, librarian, and printer was a common phenomenon after the invention of moveable type in the mid-15th century.

Concurrent with these developments in libraries during the latter part of the 19th century was the development of the social sciences. The population in the United States was growing, as were the associated social problems: increased crime, changing family relationships, abandoned children, health issues, and mandatory education. Needed were professionals to teach, manage schools, manage hospitals, and resolve the emerging social problems. The demand for professional services fostered the development of disciplines and professions to address these social problems.

With the growth of social sciences came specialists, or professions, as we know them. When the privileged classes discovered the need for special knowledge to manage emerging social problems, the professions developed, with libraries to support them.

In 1853, library leaders in the United States met to discuss the emerging field of librarianship. The successful meeting of almost 100 people led to the conclusion that they should meet again. The Civil War and other events intervened, and the next meeting did not occur until 1876. It was the centennial year of the Declaration of Independence, and associated celebrations gave impetus at that meeting to the creation of the American Library Association. Immediately, the fundamental components of a profession began taking shape through conversations about standards of service, a code of ethics, a professional journal, and other elements of a professional infrastructure.

In 1887, Melvil Dewey founded the Program of Library Economy at Columbia University. The program consisted of undergraduate courses that taught how to manage a library, including how to organize and maintain a collection of books. Pratt Institute, Syracuse University, and other professional schools were created shortly after.

Early in the 20th century, library education programs evolved from a few courses to an undergraduate major and eventually a graduate master's program. The Williamson Report (1923) urged a fifth-year degree, which began to emerge in the 1930s. Beginning in 1950, the fifth-year bachelor's degree in library science was converted to a master's degree. With the establishment of professional education programs and a structured profession, librarians were firmly established as a critical and valuable asset within the information infrastructure.

Further Reading

Because the rich history of libraries and the library profession cannot be summarized adequately in a few pages, we recommend further reading in these sources listed in the chapter bibliography: Battles (2003); Bivens-Tatum (2012); Bobinski (2007); Casson (2001); Dickson (1986); *Dictionary of American Library Biography* (1978, 1990); Greer, Grover, and Fowler (2013); Harris (1995); Johnson (1965); Lerner (2009); Murray (2009); Shera (1976); Vann (1978); Wiegand (1996, 2011); and Zeegers and Barron (2010).

Information Infrastructure Overview

The infrastructure for information transfer evolved as human knowledge evolved, and with it the technology to address human needs, including communication. While the narrative above outlines the progress in infrastructure development, Table 3.2 presents a chronology that summarizes that progress. The information infrastructure is discussed in greater depth in Chapter 4, including emergent paradigm changes in each of the information transfer processes.

Table 3.2 Transportation and Communication Infrastructure Timeline

50,000 BCE	Primitive humans draw on cave walls
6000 BCE	Egyptians use ships for transportation
3300 BCE	Sumerians write on clay tablets
2500 BCE	Ink is first used
2000 BCE	Egyptians develop postal service
1000 BCE	Chinese develop postal service
100 CE	Chinese invent paper
700	Chinese circulate government newspaper
1368	Charles V forms Royal Library; later becomes Bibliothèque Nationale
1454	Johannes Gutenberg invents printing press
1609	1st newspaper published in Germany
1636	Harvard University and its library established
1690	1st newspaper published in American colonies
1702	1st daily newspaper printed in England
1731	Benjamin Franklin establishes the Library Company of Philadelphia
1753	Benjamin Franklin named postmaster for U.S. colonies
1753	British Museum founded
1767	Iron rails used for railroad
1783	1st daily newspaper published in United States (Philadelphia)
1785	U.S. road system created
1807	Fulton launches steamboat
1814	Steam-driven press publishes London Times
1829	1st main-line U.S. railroad
1829	Louis Daguerre invents photographic process using copper plates
1835	New York passes legislation to create school district libraries
1837	Telegraph used by Great Western Railway in England
1838	1st steamship crosses Atlantic
1850s	Public libraries founded in Boston, New York, and Philadelphia.
1858	Registered mail service begins in United States
1858	Richard Dudgeon builds 1st steam-powered car
1860	Pony Express begins service
1860	Étienne Lenoir patents an internal-combustion engine
1864	Money order service begins in United States
1876	Alexander Graham Bell invents the telephone
1876	American Library Association founded
1885	Registered mail service begins in United States
1885	AT&T formed; Karl Benz builds 1st automobile
1886	1st linotype machines used to publish New York Tribune
1887	Melville Dewey founds 1st library school at Columbia University
1888	George Eastman uses paper to record photographs
1889	Thomas Edison invents motion pictures
1891	Daimler builds 1st truck powered by an internal-combustion engine
1895	Buses powered by internal-combustion engines 1st produced
1897	French are 1st to use trucks in the military
1901	Guglielmo Marconi sends radio signal between two cities
1902	Oldsmobile begins mass-production
1903	Wright Brothers' 1st flight
1904	Radio is installed in 124 ships at sea
1913	Parcel post service begins in United States
1913	Henry Ford introduces moving assembly line for car production
1914	Last horse-drawn buses stop running in London
1918	Mail 1st delivered by air
1920	1st air passenger service; 1st commercial radio broadcast
1927	Lindbergh's 1st transatlantic flight
1929	BBC begins broadcasting television

(*continued*)

Table 3.2   Transportation and Communication Infrastructure Timeline (*Continued*)

1939	Flight of 1st jet plane
1947	Cell phone invented
1950	Master's degree established as professional degree for librarians
1952	1st intercontinental jet passenger service
1953	U.S. government approves use of color television
1955	Certified mail service begins in United States
1958	1st satellite launched by U.S. Air Force
1969	U.S. Department of Defense creates ARPANET
1979	1st cell phone system initiated in Tokyo
1982	U.S. government approves use of cell phones
1983	International standards developed for the Internet
1990s	The Internet is popularized
2000–	Merging of media—Internet, phones, computers, etc.
2006–	"Cloud" storage is popularized

Public Library Services That Reflect the Emergent Paradigm

The establishment of the United States government is an example of a paradigm shift. Whereas other nations placed the authority of governance in the hands of royalty or dictators, this country's founders established a break with tradition (a paradigm shift) by placing the authority of government in the hands of the people, as outlined in the Declaration of Independence:

> We hold these truths to be self-evident, that all men are created equal, that they are endowed by their Creator with certain unalienable Rights, that among these are Life, Liberty and the pursuit of Happiness. —That to secure these rights, Governments are instituted among Men, deriving their just powers from the consent of the governed, —That whenever any Form of Government becomes destructive of these ends, it is the Right of the People to alter or to abolish it, and to institute new Government, laying its foundation on such principles and organizing its powers in such form, as to them shall seem most likely to effect their Safety and Happiness. (Declaration of Independence 1776)

This revolutionary shift of power is evidenced as well in the Preamble to the U.S. Constitution:

> We the people of the United States, in Order to form a more perfect Union, establish Justice, insure domestic Tranquility, provide for the common defence, promote the general Welfare, and secure the Blessings of Liberty to ourselves and our Posterity, do ordain and establish this Constitution for the United States of America. (Constitution of the United States 1787)

This shift was a result of the intellectual ferment of the Enlightenment and the technological ferment during the Industrial Revolution that created a large middle class, and the 19th century was a period of systemic change in the professions and the social sciences. Beginning in mid-century, the social sciences emerged in order to study people and their behavior, as the intelligentsia served on prison and hospital boards, as well as other social service agencies. From this interest in human behavior came the development of disciplines that provided the theory for social service professions. For example, John Dewey wrote treatises on education that influenced the organization of schools and curricula that remain in effect today.

Mercantile and mechanics' libraries that emerged during the mid-18th century were evidence of this new people-centered paradigm. Mercantile libraries were precursors of the public libraries that were established in Boston, New York, and Philadelphia in the 1850s.

A paradigm shift occurred in public libraries, led by such leaders as Melvil Dewey and Samuel S. Green (see "The Library Profession" above). Dewey and Green encouraged librarians to work with clientele to help them use libraries and their resources—a shift in the paradigm from archiving books and materials to provision of services to meet the needs of library users. Melvil Dewey expressed this shift in thinking in his classic work, "The Profession," published in 1876:

> The time *was* when a library was very like a museum, and a librarian was a mouser in musty books, and visitors looked with curious eyes at ancient tomes and manuscripts. The time *is* when a library is a school, and the librarian is in the highest sense a teacher, and the visitor is a reader among the books as a workman among his tools. Will any man deny to the high calling of such a librarianship the title of profession? (5–6)

By declaring that "a library is a school, and the librarian is in the highest sense a teacher," Dewey was suggesting that the librarian's role is to diffuse information, the subject of Chapter 6. And diffusion is necessary before information can be utilized, as is explained in Chapter 7.

Samuel S. Green promoted "people paradigm" thinking by advocating attention to the library user and their information needs: "A librarian should be as unwilling to allow an inquirer to leave the library with his question unanswered as a shopkeeper is to have a customer go out of his store without making a purchase" (Green 1976, 327). This emphasis on service had the effect of making the librarian a vital part of the information infrastructure and laid the foundation for the emergence of the library profession.

With a new focus on library patrons, resources were devoted to clientele of all ages in public libraries. For the first time, books were published in large numbers for children. Citizens participated in the administration of libraries by serving on governing boards. Engagement of the public in the administration of public libraries served to embolden the leadership of libraries to continue the trend toward client-centered service that is the hallmark of the emergent paradigm.

The introduction of service also introduced a paradigm shift from the bibliographic paradigm, which valued the size and organization of the library collection. With this shift in values from the collection to the library user, the vocabulary of librarians was changed, as reflected in the names assigned within the organization—for example, children's services, technical services, and reference service—during the 20th century and which remain in the professional nomenclature today.

This change in language was indicative of a change in the paradigm of professional service, but the old paradigm of maintaining a collection remained strong. The size of a collection was recognized as a symbol of quality, instead of measuring quality of service. Needs of clientele were determined intuitively (or not at all) instead of assessing needs in a systematic manner, applying systematic or scientific methods of gathering data.

Although public libraries during the 20th century advanced in their implementation of emergent paradigm thinking, they were hindered because the profession had not adopted social science theory and applied it to the organization and administration of public libraries. Throughout the history of the library profession, until recently, librarians developed "principles" to guide their tasks: principles of cataloging, assigning subject headings, reference service, etc. These principles were based on experience through practice; they were not developed through the application of psychological theory, sociological theory, management theory, or other theories from the social sciences. As social science theory was applied in research during the last decades of the 20th century, the library profession began to make greater strides in its implementation of the emergent paradigm.

An example of an emergent paradigm library system is Anythink Libraries in Adams County, Colorado. Among the client-centered services is Explore Outdoors, an outdoor classroom that enables children to explore the natural world. The library system's emergent paradigm approach is expressed in its mission statement: "We open doors for curious minds." Furthermore, the values that guide the organization are "Compassion for our customers and each other; passion for our product; eagerness to learn; optimistic attitude—we believe that anything is possible" (Anythink 2014).

School Library Services Are Examples of the Emergent Paradigm

The school library was a consequence of values shifting from the bibliographic paradigm to the people-centered or emergent paradigm in libraries and the influence of Dewey. In 1835, the New York legislature passed a law permitting voters to levy a tax to fund school libraries. By 1876, 21 states had followed New York's lead.

School libraries began as book collections without librarians—an example of the bibliographic paradigm focused on a collection. During the second half of the 20th century, school libraries shifted from the bibliographic paradigm to a learner-centered paradigm. That transition can be linked to the 1969 *Standards for School Media Programs* published jointly by the American Association of School Librarians and the Department of Audiovisual Instruction of the National Education Association (since 1971 the Association for Educational Communications and Technology—AECT). That collaboration brought a greater emphasis on the school librarian's role as teacher.

The educational role is declared clearly in the second set of standards. These were developed jointly by AASL and AECT: "Programs of media services are designed to assist learners to grow in their ability to find, generate, evaluate, and apply information that helps them to function effectively as individuals and to participate fully in society" (*Media Programs: District and School* 1975, 4).

During the last four decades, school librarians have partnered with teachers to teach information skills, as outlined in *Media Programs: District and School* and subsequent standards for school libraries. The teaching role is clearly stated in the most recent standards for school librarians, *Standards for the 21st-Century Learner* (American Association of School Librarians 2007, 3): "School librarians collaborate with others to provide instruction, learning strategies, and practice in using the essential learning skills needed in the 21st century."

The mission of the 21st-century school library program is articulated in *Empowering Learners: Guidelines for School Library Programs* (American Association of School Librarians 2009, 9): "The mission of the school library media program is to ensure that students and staff are effective users of ideas and information."

School librarians may team-teach with subject-area teachers or conduct classes on their own, teaching students to identify, evaluate, and synthesize information from multiple sources in order to complete assignments in classes for nearly all school subjects. In doing so, school librarians are teaching critical thinking and are engaged actively in the diffusion of knowledge, a process of information transfer that we consider the key role of library and information professionals in the 21st century.

Special Libraries and Services Reflect the Emergent Paradigm

Special librarians may work in medical, law, corporate, government, or other types of libraries that specialize is specific subject areas. Special librarians may also engage in entrepreneurship and independently practice their profession, contracting with individuals or organizations to provide information services.

The Special Libraries Association defines the role of information professionals:

> An Information Professional ("IP") strategically uses information in his/her job to advance the mission of the organization. This is accomplished through the development, deployment, and management of information resources and services. The IP harnesses technology as a critical tool to accomplish goals. IPs include, but are not limited to, librarians, knowledge managers, chief information officers, web developers, information brokers, and consultants. (Special Libraries Association 2014)

As suggested by this statement, special librarians customize their service to address the needs of their clientele. The emphasis is on customized service—not maintenance of a collection of resources. Customized service then becomes a tool for diffusion and utilization of information. The client is at the center of the special library; consequently, special libraries and information-broker or consultant services providing this level of service are examples of the emergent paradigm in library and information services.

Academic Libraries and the Emergent Paradigm

Libraries have had a long-standing tradition of focusing on the book and have played a valued role in the scholarly communication system as it has supported the creation of an idea into a publication. The book has been an end in and of itself, a static document or artifact. The librarian collects, stores, catalogs and classifies, circulates, and inventories books.

Feather (2004) accurately depicts the librarian's preoccupation with the book; the schemes and technological systems have functioned as a closed system, thus reinforcing the notion that the librarian's role is to purchase and protect the book. Libraries were then the warehouses provided for the storage of books. It is long-held public knowledge that the role of the librarian has been to serve as a gatekeeper of the culture's information. However, what may not be as well known is that librarians have discussed among themselves for more than a century the idea of providing service. This paradigmatic shift away from the library as a warehouse of books and toward the responsibilities of serving the needs of information users began with librarian and educator Melvil Dewey, whom we discussed earlier in this chapter. The new role for librarians is anticipating information needs and customizing (repackaging) information.

At one time, librarians focused solely on providing materials and services to college and university faculty. Today more emphasis is placed on the campus community as a whole. This includes researchers, administrators, students, staff, and faculty. In addition, academic librarians are focused on their relevance to the institutional mission. Today there is a growing interest in making the scholarship of the institution more visible and in promoting the global reach of the research and teaching that is created. As higher-education academies build institutional repositories, support publishing programs, adopt open-access policies, establish learning commons, embed librarians in academic departments, and investigate alternative metrics, the academic library continues to evolve. Academic libraries are embracing the opportunity to provide growing services to promote the impact of scholarship and programming to support this expanding mission.

This expanded mission comes at a time when providing customized information services for all constituents in the academic community is desired, and academic librarians have become a vital part of the research support needed by the academy. They have formed a partnership with the academic community through a variety of services, which include access to published research collections and bibliographic instruction—services that have added to the ease of using the vast collections of the average campus library. As this shift from the book to the user has progressed over several decades, academic librarians have grown their services into departments that focus on providing instruction, developing workflows in which some librarians are embedded within key courses. Many professors in higher-education institutions across the country now look to these information providers as co-instructors, and the services continue to flourish.

Information literacy instruction has become a prominent service for librarians due to encouragement by professional agencies like the Association of College and Research Libraries and the American Library Association, which provide professional development opportunities. The library is the logical place to support such initiatives through collaboration with faculty in teaching copyright, access to content, and publishing; this collaboration will benefit faculty as they work to develop more open course materials to counter the high cost of textbooks.

The library is truly becoming the "heart of the campus" as academic librarians further develop their roles to support their work in broad dissemination of and access to the institution's scholarship. This is in large part an action that responds to our changing society in this digital era, in which interacting with information supports a participatory culture where individuals are able to learn in groups in an online environment.

Librarians have become a vital part of the instruction for college students. Librarians are taking an active role with the instruction of information literacy skills, building and publishing local research materials and data sets, providing open-access options, and forging complex relationships with other colleagues on the academic campus. These relationships have fostered the collaboration that has contributed to the creation of the information commons. Transformations in the publishing and education delivery environments are providing abundant opportunities for collaboration and partnerships among academic librarians and those in the academic community.

Academic libraries are also creating research commons where, in designing the physical space of the actual library buildings, services and technological systems are also built to support group learning and interaction. The open access movement provides access to data and information worldwide to support research and learning. Librarians are leading in a collaborative movement to support both learning and research and to bridge both the technology divide and the participation gap, using the Internet to enhance communication.

While supporting the teaching and learning process for the education of students, a major function of the library is to partner with the faculty while they pursue research, teaching, and service. The library has been a major contributor to the process of publication and access to scholarship. It is one now because the library field has recognized that the academic library's function is to support research and teaching and to be a partner. It is as a partner that there is a move away from service and more into a collaborative role of building local content and collections of the institution's scholarship and research collections. The library has always been a silent partner, and now it is becoming a fundamental player on campus.

The university's mission has been actualized by the library as a contributing researcher for society's well-being and advancement. The academic library is a vital contributor to the university's mission.

Summary

As noted in the overview of the information infrastructure and current library and information services, it is apparent that old and emergent paradigm technologies, organizations, and services can and do function concurrently. For example, digital technologies have transformed all of the information transfer processes, yet much of the old-paradigm technologies and services may still exist. Although the computer has transformed the dissemination of information so that individuals can disseminate information instantly through the Internet, former technologies like books and periodicals still exist in their analog forms (paper books and periodicals) concurrently with eBooks and online periodicals.

As airlines transformed long-distance transportation and in some cases superseded railroads and ships, the earlier transportation modes still exist, but their function has been altered. Railroads' provision of passenger service has diminished, but their relevance to shipping goods long distances remains, although railroads serve as conveyers of truck bodies, combining truck transportation with the traditional function of rail transportation. Merging technologies are part of the emergent paradigm transformation.

Libraries have been a significant part of the information infrastructure, and all types of libraries have evolved as the digital age has emerged. This evolution is discussed further in Chapter 9. Chapter 4 will investigate in more depth the transformation that is occurring in the creation, dissemination, organization, diffusion, and preservation of information.

References

American Association of School Librarians. 2007. *Standards for the 21st-Century Learner*. Chicago: American Association of School Librarians.

American Association of School Librarians. 2009. *Empowering Learners: Guidelines for School Library Programs*. Chicago: American Association of School Librarians.

Anythink: A Revolution of Rangeview Libraries. 2014. http://www.anythinklibraries.org

Battles, Matthew. 2003. *Library: An Unquiet History*. New York: W. W. Norton.

Bivens-Tatum, Wayne. 2012. *Libraries and the Enlightenment*. Los Angeles: Library Juice Press.

Bobinski, George S. 2007. *Libraries and Librarianship: Sixty Years of Challenge and Change, 1945–2005*. Lanham, MD: Scarecrow Press.

Casson, Lionel. 2001. *Libraries in the Ancient World*. New Haven, CT: Yale University Press.

Constitution of the United States. 1787. http://www.archives.gov/exhibits/charters/constitution.html/

Declaration of Independence: A Transcription.1776. http://www.archives.gov/exhibits/charters/declaration_transcript.html/

Dent, A. A. 1972. "Before the Machine." In *Transportation Through the Ages*, ed. G. N. Georgano, 318. New York: McGraw-Hill.

Dewey, Melvil. 1876. "The Profession." *American Library Journal* 1:5–6.

Dickson, Paul. 1986. *The Library in America: A Celebration in Words and Pictures*. New York: Facts On File Publications.

Dictionary of American Library Biography. 1978. Ed. Bohdan S. Wynar. Littleton, CO: Libraries Unlimited.

Dictionary of American Library Biography. Supplement. 1990. Ed. Wayne A. Wiegand. Englewood, CO: Libraries Unlimited.

Encyclopedia Britannica. "History of publishing: first newspapers." http://www.britannica.com

Encyclopedia Britannica. "Postal system." http://www.britannica.com

Feather, J. 2004. The Information Society: A Study of Continuity and Change. 4th ed. London: Facet Publishing.

Georgano, G. N., ed. 1972. *Transportation Through the Ages*. New York: McGraw-Hill.

Grant, H. Roger. 2003. *Getting Around: Exploring Transportation History*. Malabar, FL: Krieger Publishing Co.

Green, Samuel S. 1976. "Personal Relations between Librarians and Readers. In *Landmarks of Library Literature, 1876–1976*, ed. Dianne J. Ellsworth and Norman D. Stevens, 319–330. Metuchen, NJ: Scarecrow Press.

Greer, Roger C., Robert J. Grover, and Susan G. Fowler. 2013. *Introduction to the Library and Information Professions*. 2nd edition. Santa Barbara, CA: Libraries Unlimited.

Harris, Michael H. 1995. *History of Libraries in the Western World*. Metuchen, NJ: Scarecrow Press.

Johnson, Elmer D. 1965. *A History of Libraries in the Western World*. Lanham, MD: Scarecrow Press.

Lay, M. G. 1992. *Ways of the World: A History of the World's Roads and of the Vehicles That Used Them*. New Brunswick, NJ: Rutgers University Press.

Lerner, Fred. 2009. *The Story of Libraries: From the Invention of Writing to the Computer Age*. London: Continuum.

Macintyre, Donald. 1972. "Ships." In *Transportation Through the Ages,* ed. G. N. Georgano, 121–191. New York: McGraw-Hill.

Meadow, Charles T. 2006. *Messages, Meaning, and Symbols: The Communication of Information*. Lanham, MD: Scarecrow Press.

Media Programs: District and School. 1975. Chicago: American Library Association; Washington, DC: Association for Educational Communications and Technology.

Murray, Stuart A. P. 2009. *The Library: An Illustrated History*. New York: Skyhorse; Chicago: American Library Association.

Schwartz, Peter and James A. Ogilvy. 1979. *The Emergent Paradigm: Changing Patterns of Thought and Belief*. Menlo Park, CA: SRI International.

Shera, Jesse H. 1976. *Introduction to Library Science: Basic Elements of Library Service*. Littleton, CO: Libraries Unlimited.

Smith, Anthony. 1979. *The Newspaper: An International History*. London: Thames and Hudson.

Special Libraries Association. 2014. "About Information Professionals." https://www.sla.org/career-center/about-information-professionals/

Vann, Sarah K., ed. 1978. *Melvil Dewey: His Enduring Presence in Librarianship*. Littleton, CO: Libraries Unlimited.

Wiegand, Wayne A. 1996. *Irrepressible Reformer: A Biography of Melvil Dewey*. Chicago: American Library Association.

Wiegand, Wayne. A. 2011. *Main Street Public Library: Community Places and Reading Spaces in the Heartland, 1876–1956*. Iowa City: University of Iowa Press.

Zeegers, Margaret and Deirdre Barron. 2010. *Gatekeepers of Knowledge: A Consideration of the Library, the Book and the Scholar in the Western World*. Oxford: Chandos.

4

The Information Infrastructure

Chapter Overview

In this chapter we define "information transfer" as a model for examining the information infrastructure. The processes of information transfer —creation, recording, mass production, dissemination, bibliographic control, organization by discipline, diffusion, utilization, preservation, and discarding—are described. Technology and other factors in our environment have changed these processes over time. The body's vascular system is used as a metaphor for the complex interactions within the information transfer processes, and the transportation system is presented as a way of conceptualizing the very complex information infrastructure.

Defining Information Transfer

"Information transfer" is a type of communication, as shown in Figure 4.1. It can be defined as the communication of a recorded message from one human to another. While communication assumes that the sender and receiver(s) of a message are contemporaries, information transfer requires a recorded message transmitted through a medium that enables senders to transmit ideas to people who are not their contemporaries. In other words, information transfer is asynchronous.

The recorded message can be stored in an *information system*. This system can be a library, a computer storage device, an iPod, a phone, or any kind of system that collects, organizes, stores, and makes available the information created by the sender and recorded as an information package.

However, the concept of information transfer is more complex than described above. As we consider the information infrastructure, this model describes the "life cycle" of information. The various parts of information's life are listed below:

1. Creation

2. Recording

3. Mass production

4. Dissemination

5. Bibliographic control

6. Organization by disciplines

7. Diffusion

8. Utilization

9. Preservation

10. Discarding

Following is a brief description of each of these steps or stages of the information transfer cycle. This model is explained in more detail in Greer, Grover, and Fowler (2013).

Creation

Information is created in a variety of ways. The research process that is conducted in universities, think tanks, or government agencies is a formal way for introducing information into society. Information is created by the assembly of data in ways that provide new meanings or understandings that show new relationships.

For example, researchers may conduct a formal study of the way that children react to violence portrayed on television. The researchers define a group (for example, sixth-grade boys and girls), identify exemplars of that group, and determine methods or strategies for gathering data about their reactions to a recorded violent episode. Data may be collected using questionnaires, interviews, observations, analysis of records, or other means. After the data are collected, the data are analyzed to address the questions or hypotheses that were used to frame the research process. This is an example of the beginning of a formal research process.

Another example might be the coverage of a news event by a reporter for a local newspaper or television or radio station. The editor assigns a reporter to cover a meeting of the local school board. The reporter attends the meeting, takes notes, and at the conclusion of the meeting interviews the chair of the school board and the superintendent to get their reactions to

Figure 4.1 Conventional Library Service Model

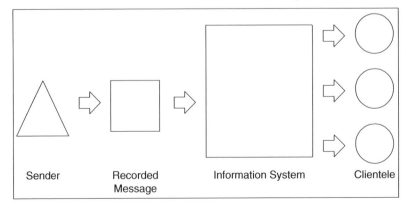

Sender   Recorded   Information System   Clientele
                Message

the key agenda items discussed and action taken. Based on the information gathered, the reporter prepares a story on a computer and e-mails it to the editor for editing and assignment to a next edition of the newspaper, which is likely to be online as well as published in paper format.

An individual, who can post his ideas on a webpage, blog post, tweet, or Facebook entry, perhaps assembling ideas from several sources and synthesizing them into a post, can also create information. Newer technologies enable an individual to create, record, and make information available for a public quickly through the use of these technologies.

Technology enables collaboration in the exchange of ideas and creation of information in ways not possible previously. For example, teams working together in a company can collaborate via software or social media platforms to create new policies, sales strategies, and services while living and working in far-flung offices or homes.

In these examples, the creation of new information was accomplished. Data were assembled into new information. However, this is only the first step in the information transfer process, and the information has little meaning until it is synthesized and recorded.

Recording

After information is created, it must be recorded in some format so that it can be disseminated and used by others. In the case of formal research, researchers collected data, and the data were analyzed to reveal trends. The analysis of data is synthesized into a report of the research process and the findings. This preparation of a report is the recording stage.

In the example of a news event, the reporter has attended a meeting and must now assemble her notes into a report of the event, a school board meeting. The audio or video recording and/or written notes from the interviews are integrated into the report.

In the third example, the creator synthesized information for recording on a Web site, blog, or Facebook post. Unlike the other two examples, there was no editor to serve as a filter for the information. The creator served both roles.

In all three examples, data that were collected have been synthesized and recorded. The recording may be accomplished digitally, using a word processor. If the reporter were working for a television station, his or her report might be recorded digitally after he or she had reported live during the local 10:00 p.m. televised news program and posted on the station's Web site. Regardless of format, the new information has been recorded as the second stage of the information transfer cycle.

Reproduction

Reproduction of information requires copying the information so that it can be distributed; however, through the use of digital technologies, mass production may be instantaneous with the dissemination. In the example of formal research, researchers send their finished report of the research project to a journal that is likely to publish a report on the topic. At this point an evaluation process occurs. A scholarly journal editor will scan the manuscript and select a group of peers to review the report, usually a "blind" review process in which the readers do not know the identity of the report author(s). If the reviewers concur that the researcher has used appropriate methods, and the topic and findings are a significant contribution to the discipline's or profession's body of knowledge, the reviewers provide a favorable evaluation, and the editor publishes the report. If one or more reviewers raise serious concerns about methodology and/or conclusions, the editor may ask the researcher to revise the manuscript to address the concerns of reviewers. If the reviewers have raised serious concerns, the editor will reject the manuscript.

When issues raised by reviewers and the editor have been satisfactorily addressed by the researcher, the manuscript may be accepted for publication. The journal may be mass-produced in paper format and/or published electronically. In either case, the review by peers is a critical part of scholarly mass production of information and knowledge.

A recent trend is the "open access" movement whereby an author or collaboration of authors submits their report to a repository, perhaps on their own campus. The repository provides open access to the report, and the authors may have the option of publishing in a journal as described above.

A newspaper or other mass-media outlet has in place a similar review process before a story is published, aired, or placed on a webpage. An editor reviews the report by the reporter or staff writer, and with the concurrence of the managing editor assigns the story to a place in the newspaper so that it can be mass-produced. A television news reporter, like a newspaper reporter, covers stories assigned by an editor. Similarly, a news director selects stories that are aired. Most newspapers and radio and television news outlets now post stories and supplementary information on their Web sites and may send brief reports via social media.

In the example of an individual who assembles information and posts an entry on a Web site, on a blog, or on social media, the mere posting of the information results in "mass production" of the information. The posting can be done quickly from one's home computer or while traveling and using a laptop or tablet computer or smart phone.

Note that scholarly and popular news media have in place an editing or content screening process. However, technological advances have reduced the cost of mass production so that desktop publishing and Internet publishing enable individuals to mass-produce vast quantities of information without benefit of the screening that has been a common feature in the past. As a result, it is much easier for unauthenticated information to be mass-produced, and one's opinions can be easily disseminated for a large audience at little cost.

It is at this reproduction stage that copyright is determined. Copyright is the legal protection granted by U.S. law to creators of publications, music, video and audio productions, and works of art. This protection is available to both published (duplicated) or unpublished works, and it gives the creator the right to reproduce the work, issue derivatives of the work, sell or lease the work, and to display the work publicly. The author or creator of a work may also give permission to others to use or reproduce the work. Copyright is an important information policy that protects those engaged in the process of creating new information in any format. Recent changes in copyright are discussed in Chapter 5.

Dissemination

After reproduction, information can be distributed to a vast audience. Technology has created new channels for the dissemination of information more quickly and more cheaply than ever before.

To decrease the rising cost of publishing scholarly journals with small circulations, an increasing number of scholarly journals are published only in electronic format. The review process can be retained, but an accepted research report can be posted on a journal Web site, accessible only to those who have subscribed, or an electronic journal can be e-mailed to subscribers. Many universities have established repositories where faculty and students can publish their work following a review process. The scholar typically retains rights to the work and may negotiate publication by a scholarly or commercial publisher.

Newspapers usually distribute via both printed and electronic formats. The distribution is fast and to a large audience, including individuals and institutions. Television news also is usually disseminated both during live newscasts and on a station Web site, which may provide additional background information to a current story or additional information that was deemed of limited interest to a mass audience.

Through the application of technology, individuals now have the ability to create, record, mass-produce, and disseminate information as never before and without the intrusion of an editor or censor. The author or creator has immediate access to a wide audience.

Bibliographic Control

The organization of information for accessing is referred to as "bibliographic control." Downs (1975, 124) defined the term as follows:

> In its broadest sense, perfect bibliographical control would mean a complete record of the existence and location of every book and of all other materials of concern to libraries. It is doubtful that we shall ever reach such a utopia.

Throughout the history of the library profession, bibliography has been central to the profession. The result of this professional focus has been the creation of such societal mechanisms as national libraries, the copyright law, and national bibliographies. In the early days of the library profession, the focus was on organization and preservation of the information sources; librarians were sometimes called "bookmen," and their knowledge base was bibliography. When the concentration of the profession was on bibliographic control, the term "bibliographic paradigm" has been used to describe this value system.

Bibliography is the conceptual foundation for library science. The dominant value of the bibliographic paradigm is with the physical document, not the user. A library that embraces the emergent paradigm must incorporate this value but also embrace the new paradigm of user-centeredness. A library that embraces both the bibliographic and user paradigms has both technical services and public services, true of nearly all libraries today.

The two types of bibliography are enumerative and subject. Enumerative bibliography is a list of everything in a defined location or scope; an example is a library's shelf list, a listing of a library's collections by the order with which they are shelved or organized. A subject bibliography is a listing of titles, regardless of format, and fulfills stated parameters defined for a purpose or a set of users. Whereas enumerative bibliography is organized by the scheme used to store and retrieve information packages, a subject bibliography is a selection of those items that address specified user needs.

In the early days of libraries, the librarian organized the collection of scrolls or books for retrieval and use. As schemes were developed for organizing books and other resources, bibliography was the intellectual foundation of the profession. When Melvil Dewey introduced the Dewey Decimal Classification System to organize books by subject for the benefit of library users, he also introduced in the 1870s the concept of service. He introduced a new service paradigm, a value system that evolved from a focus on the collection to a focus on the users of the information.

When we use the term "bibliographic paradigm," we are referring to the value system that values collections and their organization and preservation with emphasis on publishers, with less emphasis on subject, which addresses consumer access. The movement toward subjects and attaching significance to subjects was long after Dewey. That paradigm is still apparent in libraries and other information agencies in this digital age. The move from enumerative bibliography to subject bibliography was a major shift in the profession.

It should also be noted that bibliography is essential to the efficient retrieval and use of information; however, the library profession has taken a general approach to organizing information. Terms used for subject headings are standardized but are not necessarily the terms used by the library's clientele. In card and printed catalogs, limits were placed on the number of subject headings that could be used; however, advancements in computer technology have enabled the use of key words in addition to subject headings or descriptors in searching for and retrieving information. The use of sophisticated search engines provides full text searching and much greater retrieval capability.

The individual who posts on a Web site or blog might have her information accessed by a user who uses a search engine. With the advances in retrieval technology, the production of information has increased along with the sophistication of retrieval technology. Downs's assessment that total bibliographic control is impossible is valid and remains an unreachable utopia. Because of the impossibility of "controlling" information, we will refer to "bibliographic control" and "organization by discipline" as "organization of information." However, we will describe below another element of information organization, organization by discipline.

Organization by Discipline

While bibliographic control is concerned with the generic organization of information and knowledge, groups of people may design unique systems for organizing information for their use. For example, a public health agency may design a system of subject headings and a code for classifying (cataloging) their printed resources. The scheme would use an organizational scheme that would facilitate browsing and use by the clientele. Terms assigned as subject headings for the agency's catalog would be those used by the agency.

Disciplines have produced their own indexes as examples of organization by discipline. PsycINFO provides subject access to psychology research, and the Modern Language Association's *MLA International Bibliography* are two examples of the many specialized indexes and bibliographies available to provide access to the new knowledge in a discipline or profession. In the sciences, Chemical Abstracts Service (CAS), a division of the American Chemical Society, provides access to chemical research through SciFinder and STN databases.

Full-text databases enable a researcher to search a vast quantity of resources quickly, but a skilled researcher's knowledge of the field and experience using databases is necessary for a fruitful search. Organization and retrieval of information is a critical function for a society. Both bibliographic

control and organization by discipline are means for organizing information
for retrieval.

Diffusion

After information is disseminated and organized, it is the mission of
organizations and professionals in various fields to diffuse the information;
that is, to help individuals to understand this information and to make
sense of it. How do diffusion and dissemination differ? "Dissemination" is
making information available, as a newspaper makes information available
to the public. "Diffusion" is assisting in the understanding of information,
as a teacher interprets a story in a newspaper to help a child understand a
news event. History books disseminate information about world events; history teachers are engaged in diffusion of history.

Educational institutions are engaged in the diffusion of knowledge.
Public and private schools and colleges are dedicated to teaching (diffusing) basic cultural knowledge to new generations of young people. Parents,
churches, libraries, and such organizations as 4-H, Girl Scouts, and Boy
Scouts also promote the diffusion of knowledge to youth. Undergraduate
and graduate programs in colleges and universities prepare people for professions and disciplines through their educational programs. Doctoral programs prepare researchers to conduct research, creating new knowledge,
and to diffuse this knowledge to students in undergraduate and graduate
programs.

Diffusion of information is the critical function whereby people understand information so that it can be put to use. We investigate diffusion further in Chapter 6.

Utilization

The role of professions in society is to facilitate the use of information. Professionals apply knowledge to use for the betterment of humankind.
For example, medical doctors acquire knowledge of the human body, treat-
ments, and medicine, and apply that knowledge to the treatment of human
illnesses. Similarly, information professionals convert knowledge about the
creation, dissemination, organization, and utilization of knowledge to using
knowledge effectively.

Individuals must be able to take information that has been diffused
through libraries, schools, or through other channels of communication and
use that information in their lives. It is the role of information professionals
to help people understand information, to give it meaning, so that people can
use information in the personal or professional lives (see "Diffusion," above).

Use of information is the result of understanding the message received
and applying that knowledge to solve problems. We discuss facilitation of
utilization as a role of library and information professionals in Chapter 7.

Preservation

Preservation is concerned with the retention and storage of recorded
information for future audiences. When thinking about preserving records
for future users, we must think in terms of three aspects of preservation:
(1) preservation of the artifact or physical information package (e.g., book,
journal, etc.); (2) preservation of the content, the ideas; and (3) the context of
the work, its meaning at the time of its writing or production.

As with other stages in the information transfer cycle, preservation is changed by technological advances. Bibliographic control is impossible for social media, Web sites, and blogs; furthermore, preservation is complicated for organizations of all sizes. What types of e-mail should be preserved, and for how long? Which parts of a corporate Web site should be preserved? Organizations should have policies for retaining and preserving information according to the organization's mission. Often, such policies are not developed, and a great deal of information can be lost.

Discarding

Records may be removed from collections and destroyed if the content is irrelevant or obsolete, or if the physical object is damaged beyond repair. When discarding, the information professional must consider the three aspects of a record as noted in "Preservation" above. Items not preserved should be discarded, removed from a collection. Another way of thinking about this process is that resources must be evaluated regularly. When resources are no longer contributing to the mission of the agency, they should be discarded.

The information transfer cycle represents the life cycle of information, regardless of format. Electronic resources, of course, also have this life cycle. It is helpful to consider this model when thinking about the information infrastructure. Although technology has merged some of the stages of information transfer, the model is adaptable as new technologies emerge. This model will be the framework for examining the information infrastructure in this book.

The Information Transfer Processes Are Interactive

The model we have provided for information transfer appears to be linear and orderly. One process follows another, and in Figure 4.1, the information is shown to flow one way, from sender to recorded message to an information system, and to receivers of the message. However, an increasing number of library and information professionals have suggested a more interactive relationship between the information system (libraries or other information agencies) and the recipients of the information—the clientele of the information agencies. That relationship is shown in Figure 4.2.

Figure 4.2 Information Transfer Model

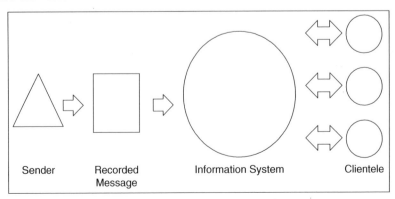

Sender   Recorded   Information System   Clientele
          Message

However, the information transfer processes are even more interactive than shown in this simplified model. We have described the information processes in a linear fashion: the creation process is followed by recording, production, dissemination, bibliographic control, organization by discipline, (hopefully) diffusion, utilization, and either preservation or discarding. Describing the information transfer process in this way makes it easier to understand, and in the orderly world of the old and still existing industrial paradigm, the information transfer process has been a step-by-step process.

For example, a newspaper reporter observed an event, recorded the event by writing a story (creation) on his or her computer (recording), and submitted it to an editor, and it was edited and published in the printed newspaper (production). The print newspaper was sent by trucks and deliverers to newsstands and home customers (dissemination). Readers read the stories, ignored some of the information, and remembered others (diffusion). Libraries collected, stored, and systematically organized the newspapers (bibliographic control), while other organizations may organize the newspapers according to a scheme they have organized (organization by discipline). Readers may act upon some of the information (utilization); then the articles deemed important might be clipped (preservation) and the remainder recycled or destroyed (discarding). This was an orderly process.

Newer technologies and information systems have changed the information transfer process so that the steps are less orderly. For example, a person might observe an event and enter it on his Facebook page (creation). As soon as it is recorded and sent, the dissemination process has occurred almost simultaneously with recording. Friends may read the entry and find it worthwhile to comment (diffusion) and even take action based on the post (utilization). Readers then may delete (discarding) or store (organization/ preservation) the post on their computers.

A major difference in the emergent paradigm world is the speed of the information transfer process, the merging of processes, and the control available to the information creator. In the old paradigm, an information creator would nearly always submit his or her work to a filtering agent, an editor or review committee, before publication/dissemination. Now, the intermediary is not present in such social media as Facebook and Twitter. As a result, inaccuracies and biases are much more likely.

Another feature of social media is instantaneous feedback from others. A Facebook entry may receive comments within minutes, and the original author of a post may respond to comments. As a result, a dialog occurs, and the creation/dissemination/diffusion/utilization processes become interactive. Similarly, most print publications such as newspapers are also online and encourage reader comments, which sometimes are published within the newspaper. The same is true in television media, where a newscast requests Twitter feedback that can result in an enhanced interactive conversation resembling positive feedback and feedforward.

The roles of information creator and consumer have been changed and become much more interactive in this digital age. As suggested in Chapter 3, the emergent paradigm dislodges the order of the old paradigm; advancements in technology have produced a variety of methods for recording, mass-producing, and disseminating information. Technology has enabled individuals to circumvent hierarchical structures, providing more freedom of expression. In the emergent paradigm, recognition is given to the notion that humans harbor a unique perspective on their perceptions of information and can more easily make their perspective known. Because the information transfer processes are now digital, the application of computer-aided technologies enables a transformation of the information transfer processes. These changes are discussed in Chapter 5.

The Blood and Vascular System Are a Metaphor

The information infrastructure is complex, constantly changing, and interactive with other systems. It is an example of the emergent paradigm. To help us explain the information infrastructure in the emergent paradigm, we likened it to the human body's vascular system: the arteries and veins that convey blood and lymph throughout the body.

The vascular system is not isolated—it is a partner with the nervous system, which is a network of nerve tissue that enables the body to react and adjust to its environment. The central and autonomic nervous systems are the two components of the nervous system. The central nervous system consists of the brain and spinal cord. The brain can be compared to a computer and the spinal cord to the cable for the computer's input and output. The nerves are like a circuit supplying information to the cable and transmitting the output to muscles and organs.

The autonomic nervous system regulates various organs of the body not subject to a human's control, including the action of the heart and blood vessels. Because it is linked to the central nervous system, the autonomic system is influenced by the emotions; for example, anger, joy, or sadness influences the rate of the heartbeat. This interconnectedness of the vascular and nervous systems of the body is like the interconnectedness of the information infrastructure.

We might compare the nervous system to a paradigm because, like a paradigm, it provides structure, meaning, and relevance as it responds to its environment. The vascular system carries the blood to all parts of the body, essentially giving it life; however, the speed of the blood's flow from the heart is regulated by the brain and nervous system. The vascular system is the piping for life-giving blood, but it is only part of an intricate network that is interrelated to the nervous system. Although the heart and brain are independent organs, they are part of this intricate network; furthermore, the brain responds to the body's environment and sends signals to the organs and vascular system. In other words, the environment influences these systems within the body.

The vascular system is responsible for continuing life, just as the information infrastructure carries information to support and enhance life. The vascular and nervous systems respond to the environment just as the information infrastructure, also a complex system, responds to the environment, including technology, policies, and culture. The information infrastructure is complex and interactive with other information-related organizations, systems, and services, and technology has enabled many changes in the information transfer processes, as described in the next section.

Technology Enhances Information Transfer

In Chapter 3 we trace the development of many technologies that support information transfer, including transportation and communication technologies. Advances in computer technologies shook the world's information infrastructure during the 20th century, especially in the recording, mass production, dissemination, diffusion, and utilization of information.

Table 4.1 Environmental Context and the Information Infrastructure

Information Transfer Processes	Environmental Context
Creation Recording Mass-Production Dissemination Organization of Information Diffusion Utilization Preservation Discarding	← Culture Geography ← Political Structure Legislation ← Economic System Technology ← Information Policy

Technology has the capability of collapsing traditional barriers to the transfer of information including distance, time, and even physical disabilities. Communication satellites provide citizens worldwide access to news events as they are happening. Cell phones enable individuals anywhere in the world to have conversations with anyone else regardless of location, and these cell phones provide a vehicle for transferring text and visual information to a wide audience through such social media as Facebook and Twitter. Furthermore, recorded information can be stored in computer files and made openly available to anyone in the world with access to computer technology. How technology has changed information transfer processes is examined in detail in Chapter 5.

Other Environmental Influences

Technology is only one of the environmental variables that influence information transfer and the information infrastructure. Others are culture, geography, political structure, legislation, economic system, and information policy, as explained in Chapter 2. These variables are external to the practice of the library and information professions, and we call them "the environmental and social context" (Greer, Grover, and Fowler 2013, 53–57).

All of the above variables influence each other and, in turn, influence each of the information transfer processes. The influences on the information infrastructure are displayed graphically in Table 4.1. The reader will note that we have combined the information transfer processes of "bibliographic control" and "organization by disciplines" and use the term "organization of information."

The environment, described as a composite of the elements in the environmental context as portrayed in Table 4.1, influences the information transfer processes and the way we think. Our environment influences the way we look at the world, and we describe that perspective as a paradigm.

In Chapter 5 we examine in depth the information infrastructure as it has evolved in the digital age.

Summary

In its most fundamental terms, information transfer is the communication of recorded information from a sender to receivers through the intermediary of information subsystems, the elements of which we define as the "information infrastructure." A dissection of information transfer reveals nine processes: creation, recording, mass production, dissemination, organization of information, diffusion, utilization, preservation, and discarding. Technology has changed these processes over time, and the digitization of information along with the introduction of a new paradigm have changed each of these processes, making them much more interactive and much less linear. The vascular system can be used as a metaphor for the complex interactions within the information transfer processes, and we used the transportation system as a way of conceptualizing the very complex information infrastructure.

Culture, geography, political structure, legislation, economics, technology, and information policy are environmental factors that influence the transfer of information and the information infrastructure. In Chapter 5 we examine the information infrastructure today and the changes that have occurred with digitization and influences of the emergent paradigm.

References

Downs, Robert B. 1975. "Problems of Bibliographical Control." In *Essays on Bibliography*, ed. Vito J. Brenni, 124–144. Metuchen, NJ: The Scarecrow Press.

Greer, Roger C., Robert J. Grover, and Susan G. Fowler. 2013. *Introduction to the Library and Information Professions*. 2nd ed. Santa Barbara, CA: Libraries Unlimited.

Merriam-Webster's Collegiate Dictionary 2004. 11th ed. Springfield, MA: Merriam-Webster.

5

How the Information Infrastructure Has Changed in the Digital Age

Chapter Overview

Paradigms are derived from a society. As described in Chapter 2, a paradigm is a set of fundamental beliefs or assumptions that provide a worldview: the way that we perceive the world and how it works. A paradigm is the soul of the information infrastructure. The information transfer processes described in Chapter 4 remain the same even as the paradigm changes, because the information transfer processes are the structure, but the way the processes operate may change.

As outlined in Chapter 3, the emerging paradigm is influenced by societal changes, and the changes, along with advancements in technology, have produced a variety of new methods for recording, mass-producing, and disseminating information. Here are characteristics of the emerging paradigm as they influence information transfer:

- Lines of authority are unclear.

- Technology has enabled individuals to circumvent hierarchical structures, providing much freedom of expression.

- With the advancement of technology and the application of digitization, visual information can be three-dimensional, opening new avenues for expressing ideas.

- The world is complex, and solutions are likewise complex; furthermore, many problems have no definite solutions, and "facts" change as new discoveries reveal new insights.

- Complex problems or issues usually have many causes that interact. Few simple solutions are possible.

- Cause and effect are not easily explained. In the emergent paradigm, recognition is given to the notion that humans harbor a unique perspective on their perceptions of information.

- Objectivity is not possible. Humans interpret their observations as influenced by the paradigms they have accepted.

- Processes are digital, applying computer-aided technologies in all of the information transfer processes. These technologies enable the transformation of the information transfer processes.

Postmodernism assumes there is no ultimate and universal truth and that knowledge is not privileged. Furthermore, the future is indeterminate. These complexities provide the emergent paradigm lens for examining the information transfer processes.

In this chapter we investigate the changing information transfer processes as the emergent paradigm influences scholarly research, the mass media, and our casual, everyday lives. As we gain an understanding of these changes, we gain a clearer understanding of the changing information infrastructure.

Creation of New Information and Knowledge

New research and new ideas have their origin in a paradigm, because paradigms provide us with an intellectual framework for thinking about issues and problems that we confront. The values within a paradigm suggest to us what is true, what is important, what research methods we should pursue, what questions we should address, and what is not relevant. The emergent paradigm presents the framework for developing questions that we may want to address and for how we will proceed in gathering information to address the pertinent questions.

The paradigm shift can be seen in the arts. Picasso's style shifted over time; in his early years, he drew women's forms that were recognizable, but in his cubist period, he portrayed women abstractly, leaving interpretation to the viewer. The artist presents a concept, and the viewer determines the meaning of a painting instead of the artist. This paradigm shift transfers the authority from painter to the viewer, and that transfer occurs in other art forms and in information transfer generally.

The New Look of Information Sources

Technology in the emergent paradigm has resulted in new formats for information; furthermore, technology and media are changing the ways we create new information. An example is Wikipedia, the online encyclopedia. The name "Wikipedia" is a combination of the words *wiki,* a technology for the collaborative creation of Web sites, and *encyclopedia.* Unlike traditional encyclopedias, for which acknowledged experts selected by an editorial board write articles, entries in Wikipedia are written collaboratively by the public. Users can identify themselves, use a pseudonym, or contribute anonymously. Similarly, readers can edit articles. Anyone with access to the Internet can author part of an article in Wikipedia; in this digital age, everyone with access to appropriate technologies has a voice in the public conversation. The result is a "self-correcting" feature of information in the emergent paradigm. The view is that knowledge functions best in an open system, where diverse viewpoints are valued, complexity is viewed positively, and mutual causality is a dynamic process of change.

Because of the way it is written, the wary reader must evaluate each article in Wikipedia carefully. The author(s) of an article may be anonymous, and their expertise may be unknown; consequently, factual errors may occur, but the self-correcting feature reduces error. An advantage of Wikipedia is access (it's free) and currency. Articles may appear in Wikipedia before they appear in more traditional information sources.

The same kind of reader participation occurs in some online scholarly journals and news sources. Most newspapers are published both online and in print; the online version on a Web page may provide space for readers' comments. Although some comments may be impertinent, the reader is able to view the reactions of other readers, sometimes revealing unknown background information or providing a sampling of public sentiment.

Formal Research Has Changed

The digital age has opened access to new formal research. With the availability of Internet access in the 1990s, access to information became available quickly and often at no cost, giving birth to the open access movement. Open access (OA) is the provision of unrestricted access via the Internet to scholarly research. Open access is frequently applied to scholarly research articles, but it is increasingly provided to dissertations, theses, and monographs.

Authors may publish in a journal and archive a copy of the research report in their university's repository for free use by the public. Another option is to publish in an open access journal, which provides free access on the publisher's Web site. Still another option is the "hybrid open access journal," which provides open access to some articles but may require a publication fee to the publisher.

New technologies enable an individual author to be an entrepreneur who can author a book, edit it, and publish it on paper and/or online. The author, using word processing, can design the book for publishing on paper and pay for the printing at a press of his or her choosing. The author then markets the book, fills orders, and sends the books to buyers. The information entrepreneur bypasses much of the traditional publishing process. A number of companies will provide consulting services, printing, and marketing for a fee. An entrepreneur can also publish a digital book for little cost through certain publishers that offer generous royalties; an author can also publish on a blog.

As Wikipedia has changed the way encyclopedias are written and distributed, so open access has changed the way new information is created in the formal research process and in the everyday world of mass communications. The creator has more control over the recording and production of his research, and the researcher's sponsoring university or other organization also has choices regarding the publication rights. In the emergent paradigm, every voice can be heard if it has connectivity. The result is a participatory culture.

Mass Media

The mass media—newspapers, magazines, radio, and television—have been transformed by the Internet so that the creators of information can instantaneously record information for transmission, with the knowledge that their audience may respond just as quickly over the Internet to the same audience. Twitter responses to articles and blogs by readers or viewers are becoming commonplace. Television news commentators almost immediately quote Twitter feeds, thus opening up the communication space.

Radio, television, and newspapers are no longer single media sources of information. These traditional media use the Internet to augment their medium; all have Web sites to supply similar information to that aired or printed. Consumer feedback is encouraged and sometimes included in telecasts, broadcasts, or newspaper articles. The consumer of information is involved in the mass media as never before.

With the input of multiple creators and the restructuring of the evaluation processes, accuracy is less certain. For example, professional editing traditionally preceded mass production of new information, but now the accuracy of information in the public realm is more difficult to discern.

Recording Information

New technologies have modified the recording and production processes of information transfer. For example, a news reporter can be present at an accident scene and report her observations live via television or radio. Similarly, a reporter might tweet an observation while observing an event. Both examples bypass the traditional reviewing process of editors.

Formal Research

Researchers may work in teams, a departure from the traditional single researcher. The team may conspire on recording results and submit findings first to their university's repository for open access by curious readers. Again, the results may be made available (mass-produced) electronically without the filtering of peer review; however, the research team, depending on the guidelines of the university's repository, may submit their research report to a scholarly journal for review and possible publication in an online journal.

The recording process that occurs simultaneously with creation enables researchers to utilize software that displays data sets for easier analysis and communication to readers. The recorded charts, graphs, or other visual information can be included in the distribution of the report.

Popular Information Sources

Similarly, news reports from the field can be merged with graphics, music, or file video for transmission within minutes of the first live broadcast. Technology enables the merging of visuals, music, and other still and motion pictures very quickly. Visuals can be modified easily; for example, a picture of a house can be modified structurally, and colors can be changed. In the emergent paradigm, nothing is fixed; once recorded, information can be modified and enhanced.

Because of technological advances, it is possible to create new information independently in one's office or home. A person can start a publishing business at home, create a Web site, or create a blog. The individual has gained independence and potential for much more participation in the creation and recording of information.

Reproduction of Information

Lines of division between the recording and reproduction of information have blurred in the digital age, as noted above in the "Creation" and "Recording" sections. Now recording of information can be instantaneous with reproduction when a reporter sends a message via social media. Storing a research report in a university repository that provides open access to that report is a merging of recording and reproduction. The lines between the recording and reproduction processes are blurred.

In the past, as described in Chapter 4, the recording and reproduction of new information were discrete and usually accompanied by a lapse of time. A reporter covered an event, returned to his office, typed a report, and submitted a paper copy to the editor. The editor then made revisions and delivered the copy to the pressroom, where the story was simultaneously converted to type and set for printing. More recently, that process was digitized, and the editor's revisions were digitally input for printing a hard-copy newspaper and posting on a Web site.

In the digital age, reproduction is on a global scale; news can be downloaded from a Web site anytime, anywhere in the world. Television and radio broadcasts, thanks to communication satellites, can transmit anywhere. British (BBC) and Arabian (Aljazeera) news organizations transmit news in the United States and worldwide, as do American networks. Multiple perspectives are available on virtually any issue.

Information consumers have access to a variety of information formats from various sources and offering differing perspectives or viewpoints. Writers and music composers provide alternatives from which the consumer may choose. For example, a variety of international books and newspapers are available online for downloading to tablet or laptop computers. Music is available in multiple formats: CDs, online, and broadcast over radio and via satellite. International boundaries have dissolved as artists, musicians, writers, and film producers from many countries make their creations available worldwide. The mass production of information in various formats for various purposes results in an overwhelming volume of information available for distribution or dissemination.

Dissemination

Technological advancements have drastically changed the dissemination of information. As the printing press in the 15th century unleashed the volume of publishing possible, computers and digitization have dramatically increased the volume of information and channels of dissemination.

As with other aspects of the emergent paradigm, the dissemination process has been decentralized and melded with reproduction. Dissemination includes the earlier methods of printing newspapers, books, journals, and other documents on paper, but those methods have been supplemented with digital documents. Newspapers, books, and periodicals still may be published in a paper format, but electronic versions are available for most titles available on paper, and the electronic format is becoming more prevalent and economical. These digital documents can be disseminated to laptop computers, tablet computers, and cell phones anywhere in the world.

We continue to see the traditional forms of dissemination: libraries, newspapers, television, radio, hardback and paperback books, oral storytelling, bulletin boards, signs, billboards, newsletters, telephones, and movies. However, these more traditional means of dissemination have been augmented with a dizzying array of new technology-assisted methods. Instantaneous transmission is also a new and common phenomenon.

Social media have infiltrated the traditional dissemination systems mentioned; for example, Twitter feeds accompany newspaper reports, television news programs, and other broadcasts to provide various perspectives along with the report of events. News media and entertainment media often invite reader response; a dialogue is encouraged in media that traditionally have supported one-way communication. Newer social media like Flickr provide pictures to accompany the verbal communication. The opportunity for multiple voices to participate simultaneously in communication is a phenomenon of the emerging paradigm.

Involvement of information consumers in the dialogue, along with the proliferation of communication channels and formats, creates a confusing deluge of information—a tidal wave of information that can overwhelm the populace. How can an information consumer assess the many perspectives and sources of information? What should one believe? Indeed, the concept "drowning in information, starved for knowledge" (Naisbitt 1982) is truer today than when it was written. The "information anxiety" described by Wurman (1989) has grown to the point that anxiety may overwhelm an individual, causing paralysis. How can society address potential information paralysis? Information professionals have a leadership role, as discussed in the diffusion process.

Bibliographic Control

Defining Bibliographic Control

In Chapter 4 we defined "bibliographic control" as the organization of information so that information users can access it. It's an inventory that a culture needs in order to have access to its past. This term has a long history in the library and information profession because bibliography, the organization of books and other materials, has been central to the profession. Bibliographic control suggests evaluation and inventory.

Bibliographic control is to books and library collections as a census is to a country's population. The concept of bibliographic control doesn't fit as well the newer forms of communication. The concept of bibliographic control is relevant for standard book and print aspects of information transfer; however, social media and other formats appearing are much more difficult to capture and classify. These newer media are digital, and while they are more ephemeral than books and journals on paper, they nevertheless transfer information.

Bibliographic Control in the Digital Age

Consequently, for this discussion we will modify the terms, because "biblio" is constraining in a digital age. As noted above, technology has enabled revolutionary changes in information transfer, and organization of information has also changed substantially. With digital documents such as eBooks, videos, e-mail messages, tweets, and blogs, the printed book is but one format for cumulated information. Instead of "bibliographic control," we will use the term "information organization."

Although technology has revolutionized the recording, production, and dissemination of information, technology also has enabled the potential for retrieving by words or ideas, as well as by author, title, and subject: access points in the bibliographic paradigm. Whereas in earlier days librarians assigned subject headings to aid retrieval, now computers can scan documents to retrieve by words or terms, and it is not necessary to categorize documents or books. The user can identify a key term in her own language and search a catalog or database. The organization has circumvented a professional; the user determines the search strategy, and a subject heading is unnecessary. Through access by terms, a combination of words, retrieval can approximate the goal of bibliographic control; that is, "a complete record of the existence and location of every book and of all other materials of concern to libraries" (Downs 1975, 124).

In the bibliographic paradigm, which was intended for printed information sources (books, periodicals, pamphlets, etc.), organizational schemes like the Dewey Decimal Classification System (DDC) and the Library of Congress Cataloging System (LC) were very effective. However, the digital age has brought a new way of organizing, as noted in the paragraphs above, and the question arises: Are the DDC and LC cataloging systems becoming obsolete?

We submit that these systems are still useful for print collections; they enable the cataloging and storage of books and journals by subject. Retrieval is simplified. However, with digital documents, such systems as DDC and LC are obsolete; eBooks and other digital documents can more easily be identified and retrieved by using keywords to retrieve documents. As an example, OCLC's WorldCat provides searching by keyword, author, and title; a search can be narrowed by adding year, type of work (fiction, nonfiction, biography, thesis/dissertation), format (article, book, musical score, etc.), and language. WorldCat also indicates which nearby libraries own a particular item, and it lists libraries worldwide that own that item.

Most libraries of all types now have online catalogs that provide access to their collections by author, title, subject, and keywords. As a result, the catalogs are available on phones and tablet computers for easy access nearly anywhere. While systems for bibliographic control have changed and more information is provided, the old systems developed in the bibliographic paradigm can still be used, augmented with the more comprehensive search capabilities available through computers and digitization. However, in the emergent paradigm, new technologies and communications systems in the digital age relegate the notion of bibliographic control to antique status.

Organization by Disciplines

Each discipline has its own dominant paradigms that indicate what's important and what isn't, and that outline major questions to be addressed. When a student majors in a subject, the initial classes are taken up with an explanation of the major paradigm and research methodologies in the field.

Whereas bibliographic control was an attempt by the library profession to organize all information using a general classification scheme, organization by discipline enables usage of the language and terms peculiar to a discipline. In the 1850s a cultural shift occurred when disciplines, fields of specialized study, identified themselves as science. First, the sciences subdivided into such areas as biology, chemistry, and physics. The social sciences followed, as a consequence of the social sciences establishing theories of human behavior. Each discipline in the social sciences and sciences developed a unique culture, using terms specific to their field. The fields developed different research methods, citation practices, and research paradigms; they developed different expectations for promotion and tenure in the university system.

The different cultures of academe and their different vocabularies required their own unique vocabularies for accessing information in their respective fields. Computers have enabled the searching of databases by using terms unique to a discipline, and documents can be searched without using a thesaurus as a guide to the vocabulary of a field.

The evolution of high-speed computers and sophisticated programming has permitted the information user to use her own vocabulary in the search process, and the librarian who in the past assigned subject headings has been bypassed. The Internet now enables a searcher to cross disciplines in a search for information, and sophisticated search engines suggest terms that can be used to locate information that might be found on Web sites, in databases, in books, or in journal articles.

The Internet has allowed the retrieval of information to reflect characteristics of the emergent paradigm—it is not fixed; there is no stability. The Internet evolves as new technologies and needs evolve. It is a process sometimes called "coevolution." Systems no longer stand alone; the complex parts interact. The complexity of the Internet reflects the complexity in the digital world. Because of this complexity, we have combined the information transfer processes "bibliographic control" and "organization by discipline" into "organization of information" to reflect the transformation of information transfer in the emerging paradigm.

Diffusion: How It Is Central in the Digital Age

In Chapter 4 we acknowledged that it is the mission of organizations and professionals in various fields to diffuse information—to help individuals understand information and to make sense of it. The ultimate purpose is to enable learning to improve one's life and to effect change. Diffusion of information is the critical function whereby people make sense of information so that it can be put to use. The transition from an emphasis on dissemination to an emphasis on diffusion is parallel to the move from the old to the emergent paradigm as well as from the bibliographic paradigm to the user-centered paradigm in libraries.

Changes in the Learning Technologies

Educational institutions engaged in the diffusion of knowledge, as well as a variety of organizations, now have a wide variety of communication channels to utilize in their educational missions. Technology applied to education has removed the barriers of distance and time; education can be offered at a distance through videoconferencing and use of the Internet and telephone for real-time and asynchronous exchange of ideas. Assignments, whether papers, oral reports, video productions, or a combination of media, can be submitted for immediate viewing or review by the instructor and students. The classroom can be simulated with students who are many miles apart from each other and the instructor; however, the effective use of these new media requires both instructors and students to think differently about the learning experience and to adapt to effective use of these media.

This new approach to learning is being used by K–12 schools, colleges and universities, churches, political parties, trade organizations, unions, libraries, professional associations, government agencies at all levels, and the private sector. All of the social media and information technologies can be used by these organizations to diffuse information.

In this digital age, education of all types is available in a wide array of formats. Once again, the student must select a learning environment and learning mode that is most convenient or most preferable for her learning style. Or the student's budget and lifestyle may steer the student to one type of learning environment over another. For example, a college student may prefer a traditional classroom environment, but the nearest college is too distant or too expensive, and the family budget may require attendance via online courses for a portion of the degree program. Similarly, a teacher may want to take graduate courses on campus during the summer, but family responsibilities may require his or her presence at home. A distance-learning class or classes would enable fulfilling family supervision responsibilities or working a summer job while participating in classes during free time.

Choices for Learners and Instructors

As with other aspects of the emergent paradigm in the digital age, the consumer has many choices in selecting educational opportunities in the marketplace. Likewise, those individuals and organizations offering educational experiences have many options for providing the programs they wish to offer. Although delivery of education once was face to face only, now the choices of educators and students are many.

Because of the myriad of choices, the importance of information professionals as intermediaries has been bolstered. Information professionals see the "big picture" of educational programs and media available. They can guide both teachers and students in the effective utilization of media appropriate for teaching and learning. The proliferation of information available for instruction requires an information specialist to help instructors use the Internet and other resources for teaching, and the information specialist can help the student to identify resources to help in the learning process and the completion of learning assignments.

New Role for Information Professionals

Whether working in a law office, a corporate headquarters, a college library, public library, museum, medical library, or school library, the information professional can work with individuals and groups to help learners identify and utilize information resources for a variety of purposes. The most valuable role of information professionals is to help individual information users identify appropriate information resources for their purposes and to help the clients make sense of the information they have found. It is this making sense of information that is the key to learning, the key to the diffusion of information. This role is vital in the digital age and is explored in more detail in Chapter 9.

This role may result in the reformation of learning organizations as we have known them. For example, we are seeing "learning commons" established in colleges and universities. These are areas that may be open for study 24 hours a day, with staff to assist students in their use of learning resources. School libraries have long established the librarian as a partner with the classroom teacher in teaching critical thinking and information literacy skills. In special libraries, information professionals also provide instruction in information identification, retrieval, and use.

Resources available to learners at all levels include print books and journals as well as electronic versions of the same. Databases are available for searching in the classroom, library, residences of students, and on tablet computers and cell phones. Open access to repositories on campus has expanded the availability of information resources and has also added to the complexity of identifying and evaluating these resources.

The complexity of the information environment and the rapid changes require partnerships among learners, information professionals, and instructors. The need for the information professional as an active partner in the teaching and learning processes has never been greater, because diffusion of information is central in the digital age. We explore diffusion further in Chapter 6.

Utilization

The role of professions in society is to facilitate the use of information. Individuals must be able to take information that has been disseminated and diffused through libraries, schools, or other channels of communication and use that information in their lives. It is the role of information professionals to help people understand information and to give it meaning and perhaps assist in interpretation so that people can use information in their personal or professional lives. In the digital age, as noted above in the "Diffusion" section, information professionals are playing a more important role in diffusion in order to enhance utilization.

The information professional must be able to diagnose the information needs of the individual or groups of information users in order to make the connection between the information and the possible uses of that information for the individual, whether acting alone or as a member of a group. As professionals initiate a diagnostic process with their client(s), they are accepting responsibility for the outcome: that the client understands the information and its possible utilization. The information professional assists individuals as they understand information and see how it can be utilized. This role is vital in the information infrastructure. A more detailed discussion of utilization is found in Chapter 7.

Preservation

Preservation is concerned with the retention and storage of recorded information for future audiences. As noted in Chapter 4, bibliographic control is impossible for social media, Web sites, and blogs; and preservation is complicated as well. The ability of individuals and organizations to create Web pages, blog posts, and e-mail messages, and to engage in the creation of social media communications (Twitter, Facebook, etc.) also complicates the preservation process.

In earlier times, the quality of paper and the extent of a document's deterioration were major factors in determining whether a document should be preserved. Duplicate copies could be procured to preserve the intellectual content of the book or document.

Digital documents create a new set of issues. The ephemeral nature of social media is a factor in the preservation process, and individuals must make decisions regarding the value of their communications. Those decisions can be made on an ad hoc basis. However, digital communications for organizations can be very important for decision-making, for policy-making, and for legal protection. Web sites, digital documents, and e-mail should be included in the retention and preservation policies of an organization.

Unlike paper documents, digital documents more easily can be stored without regard to storage capacity; however, the format of the digital document must be considered. As hardware changes are made in the organization, capability for retrieving digital documents in the organization's library is a consideration in preservation policy. Another consideration is the protection of digital documents by systematic storage and organization so that the documents are not lost within the organization and accidentally deleted. As more documents and communications are digitized, individuals and organizations must be intentional about developing organizational, preservation, and discarding policies.

Discarding

The process of discarding digital records is much the same as with paper documents. Items may be discarded if the content is irrelevant or obsolete, or if the digital document is flawed or cannot be retrieved because of obsolescence. Resources in all formats should be evaluated regularly. When resources are no longer contributing to the mission of the agency, they should be discarded. Policies should be developed to provide for a regular and systematic review of documents held by an organization.

The Information Infrastructure in the Digital Age

As we described the information transfer processes, the building blocks of the information infrastructure, we noted how the divisions between the processes or stages of information transfer were no longer discrete. They overlap each other, yet these processes are all influenced by the environmental context that is outlined in Chapter 4: culture, environment, political structure, legislation, economic system, technology, and information policy.

A graphic model of the information infrastructure in the emerging paradigm is shown in Figure 5.1.

Summary

A paradigm is a set of fundamental beliefs or assumptions that provide a worldview: the way that we perceive the world and how it works. A paradigm is the soul of the information infrastructure. The information transfer processes described in Chapter 4 remain the same as the new paradigm emerges, because the information transfer processes are the structure, but the way the processes operate may change.

Advancements in technology have produced new methods for recording, mass-producing, disseminating, diffusing, utilizing, and preserving information. In the emerging paradigm world, the central role of information professionals in the digital age is to support diffusion and utilization of information and knowledge.

Figure 5.1   Information Infrastructure

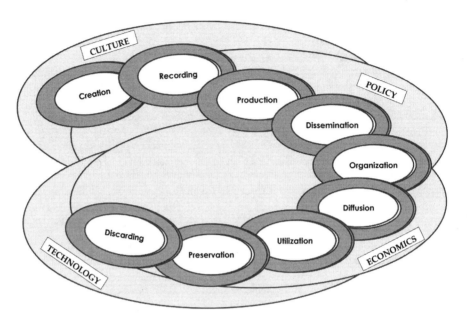

References

Downs, Robert B. 1975. "Problems of Bibliographical Control." In *Essays on Bibliography*, ed. and comp. Vito J. Brenni, 124–144. Metuchen, NJ: The Scarecrow Press.

Naisbitt, John. 1982. *Megatrends: Ten New Directions Transforming Our Lives*. New York: Warner Books.

Wikipedia: The Free Encyclopedia. http://en.wikipedia.org/

WorldCat. http://www.worldcat.org/

Wurman, Richard Saul. 1989. *Information Anxiety: What to Do When Information Doesn't Tell You What You Need to Know*. New York: Doubleday.

6

Diffusion of Information and Knowledge: How Information Becomes Meaningful

Chapter Overview

The purpose of libraries is the diffusion of knowledge. Over time libraries have shifted from a concentration on the identification, acquisition, storage, access, and preservation of information, a passive mode of services, to a service-centered mode with more interaction with information users. In a global market-oriented world, diffusion and utilization of knowledge and proactive services to deliver that knowledge are critical to society. Diffusion of knowledge is a key element in information transfer and in the information infrastructure. Diffusion refers to the flow of knowledge from a source to the adopter. It denotes a process of accepting the message, processing it, and integrating it with existing knowledge; it is social learning, embracing the spread of a new idea within an individual or social system.

The Difference between Dissemination and Diffusion

While the library and information science and social science literature is rich with differing definitions of "dissemination" and "diffusion," much confusion exists with the definitions and use of these terms. The terms are often used interchangeably, and only through a close examination of these terms and definitions can one have a clearer understanding of the intended meaning.

Dissemination

Dissemination means simply the distribution of products or information. Everett Rogers (2003) states that the goal of diffusion of innovation theory is to explain how innovation—whether ideas, products, or processes—is shared. Dissemination can be conceptualized as introducing new research or new ideas into the public domain through broadcast or publication in one or more means, as suggested below. An early effort to define dissemination is by Klein and Gwaltney (1991, 246–247). Their model identifies four elements in dissemination:

1. Spread: the one-way broadcasting of information in order to increase awareness, that is, "proactive dissemination." Examples include databases, monographs, newsletters, newspapers, information packages, and research reports.

2. Choice: providing information "of alternative databases, helping users acquire information they seek: reactive dissemination." Examples include retrieval services, training in use of databases, workshops, and identification of best practices.

3. Exchange: "multi-way flow of information—interactive dissemination." Examples include conferences, advisory boards, facilitators, and needs assessment.

4. Implementation: "utilization to achieve change in attitudes or behavior." Examples include technical assistance, training, and support services.

A close examination of Klein and Gwaltney's model reveals a number of assumptions about knowledge transfer. Spread and choice are viewed as linear mechanical processes of "transfer." Knowledge is packaged and moved from one place to another. The process is linear, pushing information out with the emphasis on distribution rather than the user seeking information. For example, scientists engage in research and publish in journals that may be acquired by the Library of Congress, regional libraries, and institutions so that the research is made accessible to scholars and the public.

In the next stages, exchange and implementation, elements of diffusion are present. Change of "attitudes or behavior" implies that the message not only was received but also was understood, processed, and integrated with previous knowledge (Klein and Gwaltney 1991, 248).

As noted in chapters 4 and 5 of this book, technology has created new channels for the dissemination of information more quickly and more inexpensively than ever before. Next we will examine the importance of diffusion in the information transfer process and the role of information professionals.

Diffusion

After information is disseminated and organized, it is the mission of organizations and professionals in various fields to diffuse the information, that is, to help individuals understand this information, to make sense of it so that it can be utilized. Examples of professions engaged in diffusion include teachers, journalists, clergy, politicians, corporate sales personnel, trainers, consultants, managers, marketing specialists, and every other profession, because professions are in the business of diffusing the culture of that profession—what's true, what's acceptable, and what's not acceptable.

Occupations, like professions, diffuse the knowledge and skills of the occupation through apprenticeships, on-the-job training, and coursework.

Much of that occupational knowledge is at a skills level: how to do something in an acceptable manner. Unions may be involved in this diffusion process.

How do diffusion and dissemination differ? "Dissemination" is making information available, as a newspaper, database, or library makes information available to the public. "Diffusion" is assisting in the understanding of information, as a teacher interprets a story in a newspaper to help a child understand a news event, or a librarian helps an adult locate a book or online article to provide background on a news event, helping the individual to make sense of a newspaper article or database entry. History books disseminate information about world events; history teachers are engaged in diffusion of history through their instruction.

If we take a broader, more theoretical perspective, we might note that dissemination is an old paradigm in the bibliographic paradigm of the library and information professions; the focus is on the creation, mass-production, organization, and dissemination of information resources. The bibliographic paradigm is most concerned with identification, acquisition, organization, retrieval, and circulation of information. The emphasis is on efficiency, correctness, and thoroughness in the management of libraries and information systems. The emphasis is on counting and quantitative analysis as opposed to qualitative analysis of service to the information user.

The objective of diffusion is to provide systems for organizing, storing, retrieving, and disseminating information efficiently. Most importantly, the objective is to assist the information consumer in making sense of that information for effective use.

Melvil Dewey at Columbia University in the 1870s was the first librarian to articulate the need for service in the profession; hence, he was the first librarian to implement new user-paradigm views. Identifying the need to provide reference service to the library's clientele, he hired two reference librarians for that role. By doing so, he demonstrated a shift in his value system, the initial move toward the user as most important in management decision-making.

The values of the profession began to change, and the movement to a user-centered paradigm began. As time passed, more user services were developed, and research was conducted on these services in order to assess their value and to improve them. When the focus shifted from information sources and circulation to serving clientele, the profession more frequently applied social science theories to better understand the individuals and groups that were information users. That change in emphasis has been noteworthy in the gradually increasing application of social science theories to LIS research; examination of human behavior, learning theory, cognitive styles, role theory, social psychology, and sociology is applicable to the LIS field in order to serve clientele and to facilitate diffusion.

Diffusion of information is the critical function whereby people understand information so that it can be put to use. In Chapter 4 we acknowledged that it is the mission of organizations and professionals in various fields to diffuse information. The ultimate purpose is to enable learning to improve one's life and to effect change.

Information professionals can work with individuals and groups to help learners identify and utilize information resources for a variety of purposes. The role of the information professional is to assist individual information users to identify appropriate information resources for their purposes and to help clients make sense of the information they have found. Making sense of information is the key to learning, the key to the diffusion of information.

How Information Becomes Personal Knowledge

As we consider how learning takes place, we will consider this issue from two perspectives: that of the individual learner and that of groups. The latter is concerned with both small groups (for example, school classes) and sub-societies (for example, senior citizens).

Individual Learners

The goal of information professionals is to create and operate information systems and services that accommodate the information needs and behavioral characteristics of a specific client population. This goal of customizing information service requires knowledge of human behavior associated with the acquisition and use of information. Information psychology is the field of information science concerned with the development of this theory of human behavior. Information psychology applies behavioral theory to the library and information professions in the same way that educational psychology is an application of psychology to the teaching profession.

The authors provided an overview of information psychology in an earlier work (Greer, Grover, and Fowler 2013), and we highlight the essential ideas here. The purpose of information psychology is to address the following questions:

1. How does a person decide that there is need for information?

2. What processes are involved in decisions to satisfy or ignore the need?

3. What are strategies employed in searching for information, and how does this differ among individuals?

4. What are the variations in behaviors associated with the search for information?

5. How does the medium of the information (book, electronic publication, video, etc.) influence the individual's selection and use of the information?

6. What methods and criteria are employed in evaluating the relevance of information acquired?

7. How does information become knowledge? How does format and system design affect this aspect of behavior?

8. What cognitive styles are employed in information processing? How does learning style influence information searching, information retrieval, and information use?

9. How do individuals organize, store, and retrieve information from memory?

10. What are the varieties of forms and patterns of information utilization?

Contributing to an understanding of these questions are the research and theory from the following fields: psychology, psycholinguistics, physiology, religion, educational psychology, social psychology, sociology of information, information engineering, and information organization management.

Information psychology provides the background knowledge that enables a library and information professional to conduct an effective diagnosis or information needs assessment. Information psychology provides the knowledge of how individuals use information. Sociology of information provides insight from the perspective of an individual's membership in a group.

The information organization management and information engineering fields relate to the information processes—what information professionals *do* with information. These two areas of applied theory are unique to the practice of library and information professionals.

Role of the Professional

A professional possesses a particular knowledge that enables that individual to apply that professional knowledge with a service as the product. The role of any professional (e.g., physician, attorney, librarian, teacher, architect, or financial planner) is that of diagnosing needs, prescribing a service that meets those needs, implementing that service, and evaluating the outcome of this interaction. In most professions, this process is accomplished at two levels: with individuals and with groups, as indicated in Table 6.1 below.

This process, which we call "the service cycle," is described below as it applies to an information professional and is based on the medical model for diagnosis. The reader is urged not to dwell on the different terms used to distinguish the diagnostic processes for individuals and groups. The different terminology is used merely to emphasize the distinction between individual and group services.

Diagnosis/Analysis

The professional must be able to assess the information needs of clientele at two levels: (1) analyzing the characteristics of the community served, and (2) analyzing the needs of specific individuals at the point when and where they seek information from the library or information center. The purpose of the first level of analysis (i.e., community analysis) is to provide the professional, as manager, with specific data about the community and its residents. Knowledge acquired through a systematic process of data collection and analysis enables the information professional to understand the environment in which the library media center will operate. The school, college, or corporation, too, is a community that must be analyzed to determine the needs of the various groups that make up the community for which information service is planned. An understanding of the community and the school will provide a conceptual framework for customizing collections, services, and space allocations. This level of analysis is a first critical step in customizing library media service for individuals and groups within the

Table 6.1 The Diagnostic Process

For Individuals	For Groups
Diagnosis	Analysis
Prescription	Recommendation
Treatment	Implementation
Evaluation	Evaluation

school, and a great deal of literature exists on the assessment of user needs from the group perspective. Community needs assessment is discussed in more detail later in this chapter.

The second level of analysis is the one-on-one interaction with a user at the point when the decision has been made to seek information. The professional must diagnose the information needs of the user as the first step in the professional/client interaction. At this point, the professional initiates a diagnostic process with the client and accepts responsibility for the outcome. This interaction must begin with the basic questions of "what, why, how, when, and where" and narrow to match the professional's perceptions of such client characteristics as level of literacy, cognitive style, and social construction of reality. Once this level of needs assessment is completed, the professional will proceed to the next stage of the service cycle, that is, prescribing or recommending the source or sources from which the appropriate information may be acquired.

Prescription/Recommendation

The professional, in a one-on-one relationship with a client, will prescribe appropriate information sources in which the desired information may be located to satisfy the diagnosed need. The professional as a manager approaches the diagnosis/prescription process from an organization perspective. That is, the process is not intended to serve the needs of a single person, but rather the entire population within the library service area. The needs assessment is an analysis of aggregate data about the population of the service area and is used to create an organization customized in its design, collections, and services to fit the characteristics, behaviors, and idiosyncrasies of that population. Conclusions from this analysis can lead to informed decisions about such specifics as the size and scope of collections and services.

Treatment/Implementation

The "treatment" or "implementation" is the organization and application of the service that has been prescribed or recommended. At the individual level of service, the treatment brings the client and the needed information together. This service requires knowledge of various information sources and services that are available within the system, as well as those located elsewhere.

With the advancement of more complex and sophisticated technology, the library and information professional must be aware of (1) the array of information sources available, (2) the "best use" of a particular information package for meeting client needs, (3) the preferred formats of the client, and (4) the information needs of the client. This phase of professional service relies heavily on the diagnosis in order to determine client preferences and information needs. As a manager, the library and information professional organizes a service that addresses the information needs of a group, employing knowledge of the group's characteristics to provide the information, staff, and facilities to offer the service.

At this stage of implementation, the diffusion of information can take place. Simply presenting information sources to a client or group of clients is the dissemination of information. Diffusion requires the information professional to go a step further: the information professional must interact with the individual, either alone or in a group, to help the client understand the meaning of the information and to help the client make sense of the information. With meaning comes diffusion of information, making utilization possible. More about diffusion is found later in this chapter, and utilization is discussed in Chapter 7.

Evaluation

After the information service has been implemented, the outcome must be evaluated in terms of clientele satisfaction. An unsatisfactory resolution of the original need should trigger a repetition of the entire cycle. The second cycle may amend a part of the sequence, or it may require an entirely new approach. In a reference situation, for example, the information professional would observe and query the client after presenting information to assess the appropriateness of the information provided. Likewise, a service should be evaluated and modified according to the findings of an evaluative process. Similarly, allocation of organization resources for specific purposes should be evaluated after implementation. Methods for collecting and analyzing data for purposes of evaluation can range from simple verbal inquiries to sophisticated quantitative and qualitative analyses, depending on the circumstances.

Whether applying this process of prescription or analysis, diagnosis or recommendation, treatment or implementation, and evaluation to an individual or group, a critical component is the diagnosis of the individual's or group's need, preferences, and cognitive styles. Information psychology helps us to understand the unique aspects of diagnosing the information needs of individuals.

An Overview of Information Psychology

The model for information user behaviors is found in the framework of information psychology, as noted above. The information professional may be called upon to diagnose information need at any of the following points in the behaviors of acquiring information: awareness of need, action decision, strategies for search, behaviors in search, evaluation, assimilation, memory, and utilization. Following is a brief description of these actions:

Awareness of Need

A client determines a need for information. The need may be for informational, educational, recreational, decision-making, or research purposes. The first stage in the information use behavioral process is becoming aware of a need for information.

For example, a child is given a dog by her parents, with the condition that the dog must be well-mannered and must stay off the chairs and sofa. Shortly after the dog arrives at his new home, he jumps onto the sofa in the living room. Seeing this violation of the "furniture rule," the young girl sees that she must train the dog. She has a need for information and becomes aware of that need.

Action Decision

After becoming aware of a need, a client may elect to act by seeking to satisfy that need. In our example, the girl realizes that she must change the dog's behavior, or the dog will continue to violate the rule that he must stay off the sofa. She decides to seek information on dog training.

Strategies for Search

Once a decision is made to search for information, clients will employ their unique strategies for locating information. A plan of action is formulated, usually very informally.

Jane, the girl in our example, begins her search for information. She first consults the encyclopedia set in the home, but she finds little information about dog training. As she contemplates her choices, she thinks about other sources of information: (1) she could ask her parents for help; (2) she could call her friend Stephanie, who has had a dog for more than a year; (3) she could go to the library to find a book on dog training; (4) she could go to her home computer and conduct a search for information; (5) she could call the veterinarian that gave the dog a physical examination the day after the family acquired their pet.

Behaviors in Search

The client enacts the search strategies. These behaviors might include a Google search, consulting indexes, consulting with a librarian, asking a friend, etc.

Our young dog owner talks with her parents about her concern that the dog is misbehaving and asks if they have suggestions. The parents suggest that Jane and the dog enroll in a training class provided by the local humane society. The parents suggest that she call the local office to learn when the next class will start.

After dinner that evening, Jane calls her friend Stephanie to ask her about dog training. Stephanie invites her over after school the following day to talk about Stephanie's experiences with their dog.

Jane then goes to the computer in her home to check the electronic catalog of the local public library. She finds the titles of three books about dog training. She also conducts a Google search using the terms "dog training" and finds an excellent Web site with training videos. She bookmarks the Web site and vows to return when she has more time.

Evaluation

The client evaluates the results of the search to determine if the search should be modified, if the search should continue, or if the search should be terminated. After completing her computer search, Jane decides that several sources will be helpful, and she will not need to contact the family's veterinarian. After checking the school library for appropriate titles, she will stop by the library on the way home from school. When she gets home, Jane calls the local humane society and learns that the next training class begins in two weeks and a parent must accompany her.

Assimilation

If the results of the search are deemed satisfactory, the information may be assimilated into the client's information system; for example, the information may be copied for further use by taking notes or downloading. As the days pass, Jane reads the book she found in the school library and one of the books from the public library. She creates a document in her home computer and makes notes about training her dog. She copies and pastes into that document pertinent paragraphs from the Web site she found.

Memory

If the information is pertinent, the client may memorize the information; therefore, the information may be learned, becoming a part of the individual's knowledge. Jane has learned several tips from the sources she has read, and she enrolls in the dog-training course with her mother. They practice the dog training techniques taught in the class and take home a pamphlet that includes much of the information taught in the class.

Utilization

If learned, the information may cause behavioral changes that cause the client to use the information. Jane began to apply the information from the books and Web site immediately after reading, and she utilized class information immediately because the instructor included training exercises in the class. As the weeks passed, Jane was able to teach her dog obedience so that he soon learned to stay off the family's chairs and sofa.

At any of the stages above, the client may consult with an information professional in order to solicit help in information seeking. The professional must identify the stage of information seeking as part of the diagnostic process, using knowledge based on the theories that make up the professional's knowledge base in order to complete the diagnosis.

Summary of Information Psychology

The process whereby an individual processes information so that it becomes personal knowledge is summarized in Table 6.2. As an individual proceeds through the information-seeking processes, the individual is influenced by external forces as described in the column labeled "environmental factors." These variables are described in Chapter 4.

As the reader considers her or his own information processing, one realizes that information processing is not linear. We sometimes go back to one or more processes preceding the one we currently are on. For example, when making action decisions, we may review our need for information to clarify our need, in order to clarify our search strategies. Similarly, when we evaluate information found during a search, we may revise our strategies and seek additional types of information. The information search process requires backtracking, evaluation, and revision throughout the process.

Table 6.2 Human Information Processing

Information-Seeking Behaviors	Environmental Influences
Awareness of need	Culture
Action decision	Environment
Strategies for search	Political structure
Evaluation of information	Legislation
Assimilation	Economic system
Memory	Information policy
Utilization	Technology

How Information Becomes Social Knowledge

In the previous section, we described the process whereby an individual accesses information, assimilates it, and puts it to use. At a societal level, including small groups, organizations, professions, occupations, and societies of various sizes, the social process of creating, disseminating, diffusing, and utilizing information is a process we call "information transfer." We have described that process in chapters 3, 4, and 5. Here we focus our attention on the diffusion process.

Rogers (2003, 11) defines "diffusion" as the process by which (1) an innovation (2) is communicated through certain channels (3) over time (4) among members of a social system. Diffusion is a unique type of communication with a focus on the spread and understanding of a message or new idea.

Innovation

Innovation is the degree to which a new idea is perceived as better than the idea it supersedes. The key element is perceived advantage to an individual or members of a social system. Attributes of innovation are relative advantage, compatibility, complexity, trialability, and observability.

Relative advantage is the degree to which an innovation is perceived as better than the idea it supersedes. Advantage can be economic, social, or simply a matter of convenience. What is critical is that it is perceived to be advantageous. Perceived advantage positively affects the rate of adoption of an innovation.

Compatibility "is the degree to which an innovation is perceived as being consistent with existing values, past experiences, and needs of potential adopters." Ideas must fit the values and norms of a social system; if they do not, the adoption of an innovation or new information takes much longer.

Complexity "is the degree to which an innovation is perceived as difficult to understand and use." If a new idea is readily understood, adoption is more easily accomplished than if a new skill set is required for the adoption process. Interventions, such as workshops, become necessary.

Trialability "is the degree to which an innovation may be experimented with on a limited basis." New ideas can be tried on a step-by-step basis to help in the adoption process. Learning by doing reduces uncertainty.

Observability "is the degree to which the results of an innovation are visible to others." It reduces the risks for late adopters. Individuals can evaluate peers' success in adopting an innovation (Rogers 2003, 14–15).

This model suggests that diffusion of innovation is more likely to succeed if individuals perceive an advantage in adoption of a new idea or practice. Compatibility, trialability, observability, and reduced complexity further the process of diffusing innovation.

Communication Channels

According to Rogers (2003, 5), communication "is a process in which participants create and share information with one another in order to reach a mutual understanding." A communication channel is a means of getting a message from the sender to the user. Both mass-media channels and interpersonal channels are important in creating awareness of new ideas. Peers are of particular significance in the adoption of innovation processes, since they act as subjective evaluators.

Time

The next element in the diffusion of new ideas is time. Time is an inimical part of the innovation-decision process. The process involves a thought process whereby an individual becomes aware of an innovation, develops a positive attitude toward the innovation, makes a decision to adopt or reject implementation of a new idea, and finally accepts the decision. The innovation-decision process aims to reduce uncertainty in the adoption and the resultant consequences of a decision.

Adopter Categories

According to Rogers, there are five adopter categories: (1) innovators, (2) early adopters, (3) early majority, (4) late majority, and (5) laggards.

- Innovators must cope with a high degree of uncertainty about an innovation.

- Early adopters serve as role models and mentors in the innovation process for members of the social system.

- The early majority adopts the innovation just before the other members of a social system.

- The late majority adopts a new idea because of strong peer pressure. They are cautious and only adopt a new idea when the majority has done so.

- Laggards have a localized outlook, are wedded to the past, and are fearful of the innovation. They look for certainty that the new idea will not fail.

As noted earlier, time is linked to the rate of adoption. It is measured by the number of members of the social system that adopt the innovation.

Social System

The last element in the diffusion of new ideas is the social system. Rogers defines a social system as a set of interrelated units that are engaged in joint problem-solving to accomplish a common goal (Rogers 2003, 3). Social systems are composed of individuals, informal groups, organizations, and subsystems. Social systems define how widely innovation diffuses.

Successful diffusion systems have many common elements:

- Understand user characteristics (information needs analysis)

- Use a level of language understood by potential users

- Are accessible, available, and adaptable

- Are timely and comprehensive

- Use accepted and validated knowledge

- Use relevant materials that meet users' needs

- Use all forms of communication—print, electronic, visual, audio, person-to-person

- Use ongoing feedback and feedforward
- Are integrated with evaluated and validated research
- Use formal and informal networks
- Provide mentoring, training, technical assistance, and continuous interaction to users

Diffusion Models

Three innovation diffusion models—the interactive model, the linked-chain model, and the emergent model—describe the knowledge each model produces.

- The interactive model is a linear and sequential model of innovation with a need-pull and technological push. Internal and external knowledge is used to solve problems. The model uses the technological pull and push to realize diffusion. Market needs encourage technological innovation; at the same time, technology allows reaching new markets.

- The linked-chain model consists of six elements that are linked together: (1) market finding, (2) analytical design, (3) development, (4) production, (5) marketing, and (6) research and knowledge. Knowledge, in this model, is the core element that fuels diffusion. The steps include development of a (new) product, design of the product, testing, development of a prototype, distribution of the product, and research. This chain represents a conventional step-by-step path to innovation.

- In the emergent model, innovation is brought about through the proliferation of ideas, by competition among different potential solutions. Shock, surprise, and setbacks in the adoption of new ideas or products are beneficial. (Baskerville and Pries-Heje 2001, 183–186)

The first two models are conventional in their linearity to the problem-solution approach. The emergent model takes into account the dynamic and ever-changing reality institutions face and the need to become a learning organization.

In summary, effective diffusion is dependent on users, information sources, content of the message, medium of the message, and context:

- User groups or potential users are clearly defined.
- Information sources are highly knowledgeable in the subject area.
- Content is understood through examples, stories, and demonstrations by the users.
- The medium of the message is described, packaged, repackaged, and transmitted in a format desired by the users.
- Context for the use of new knowledge is affected by the environment, people, and support needed to facilitate use.

Influence of Culture on Diffusion

Classical diffusion studies focused on the novelty of innovation, the communication process and channels, and the adoption of innovation.

Innovation implies forward movement and change. However, since innovation is risky and uncertain, adoption is a slow process.

Katz, Levin, and Hamilton (1963, 240) added the cultural dimension to Rogers's definition of diffusion: the (1) acceptance (2) over time (3) of some specific item—an idea or practice, (4) by individuals, groups, or adopting units, linked (5) to specific channels of communication, (6) to a social structure, and (7) to a given system of values, or culture.

This model attempts to be more sensitive to specific cultural, structural, and individual effects of intercultural diffusion. Culture denotes common beliefs, assumptions, and values that guide behavior. Culture is value-laden and identifies both positive and negative beliefs, assumptions, and values.

In a global society, the diffusion of professional practice can have its limitations. Negotiating standards and policy must account for differences. In the library field, the Western notions of "service" that value the user and user input can be quite alien to librarians in developing countries.

The Western notion of service is found in the United States Constitution, clearly an American invention that has influenced virtually everything this country has done, including laws and policies. While we stray from it occasionally, we're a people-oriented society, placing the people and collective governance over authoritarian rule. Despite the democratic ideals articulated in the Constitution, ideological disputes occur regarding the limits of the government's role regarding such issues as assistance to the needy, health care, regulation of air quality, and regulations of private enterprise. The boundaries of the paradigm are dynamic and constantly challenged.

Within the library and information professions also, we find both the user-centered service paradigm as well as the bibliographic paradigm that values collections of resources and order perhaps more than the needs of users. The presence of both paradigms may be apparent in cataloging standards, access to information, and outreach efforts. The professions continuously confront the function of new technologies and their utilization within the service-oriented paradigm.

Education in the Information Infrastructure

A vital component of the information infrastructure is education, which is all about the diffusion of knowledge. And education is the diffusion of culture; the purpose of a country's or organization's education system is to teach children and adults the core culture of that country or organization— the values associated with its history and the necessary skills for a person to succeed in contemporary society.

The Role of Education

A nation's culture includes the history, music, language, literature, art, folklore, and the values held dear by that society. The educational system in the United States has its roots in Europe, dating back in time to ancient Greece and Rome. Socrates, Plato, and Aristotle provided the intellectual foundation for the nation and its educational theory. These classic ideas were studied in Europe through the Middle Ages and during the time of the American Revolution, and its leaders were schooled in classical studies that influenced the democratization of the new country. The United States Constitution reflects the changing values from monarchy and authoritarianism to a government by and for the people, a democracy.

This determination to establish a democracy required a system of public education that was a departure from the education of aristocratic men that had been the dominant trend from antiquity until the 19th century. A democracy requires the education of all members of a society, regardless of gender, socioeconomic status, ethnic origin, or religious beliefs. This government for and by the people required a new form of education.

Colonial Schools and Colleges

During the colonial period of the United States, public education was a mixed bag of schools with some of the traits of the English school system that was differentiated by class. Pulliam and Van Patten (2007, 109) note that in 1775, just before the revolution, 37 newspapers were in circulation among the colonies, helping to diffuse the culture. City, private, and college libraries throughout the colonies aided this diffusion.

Several colleges were founded during the revolutionary period, including Harvard, William and Mary, Yale, and the College of New Jersey, founded between 1636 and 1746. King's College, the College of Philadelphia, the College of Rhode Island, Queen's College, and Dartmouth were founded between 1750 and 1776. Elementary schools before the revolution were scattered, and some towns had no schools. Secondary academies and some private secondary schools were found in larger towns during the decades before the revolution, but education for all children was not part of the disorganized education systems of the colonies.

Schools in the New Nation

During the Revolutionary War, people's energies were directed away from education, and illiteracy increased because rural schools and even schools in larger towns were closed. Higher education was restricted, and books were scarce because printers could not maintain their presses without supplies from England (Pulliam and Van Patten 2007, 114).

The nation's founders clearly held the view that knowledge was essential for a free people; they did not believe that the federal government should control education. The U.S. Constitution contains no reference to education, although the First and Tenth amendments ensure nonsectarian public schools under state control.

In 1785 Congress passed the Ordinance of 1785, which specified laying out townships in 36 sections, and one section in each township in the territory was to be reserved for schools within the township. Another ordinance in 1787 stated that each new state was authorized to reserve up to two townships for a university. Later, most new states received federal land for educational purposes. As new states entered the Union through the years following the Revolution, their state constitutions all set up systems of schools.

The school movement received a boost from the Industrial Revolution, which originated in Europe and was spurred by such inventions as Watt's steam engine, Eli Whitney's cotton gin, and Bessemer's process for steelmaking. During the 1800s the American economy shifted from agriculture to industry; in the Northeast and Midwest, textile mills, coal mining, iron and steel plants, railroads, and other commercial enterprises transformed agricultural communities into mill towns connected by railroads. Industrialization caused the growth of cities, and with that growth came slums, pollution, abuse of labor, poverty, and social unrest, as well as the recognition that education was needed to supply trained workers for the emerging industrial economy.

Early in the 19th century, pressure was generated for the improvement of public schools, because workers in cities could not afford to send their children to private schools. Furthermore, the westward expansion of the United States also saw the expansion of the concept that one person was as good as another, without regard to wealth or social status. During the first quarter of the 19th century, many states began to develop school systems that still exist today.

Foundation for Modern Schools and Colleges

Permanent funds to support schools were created by several states, including Connecticut, New York, Virginia, and Pennsylvania. Acts were passed by states with a provision that local school districts might levy taxes to support schools if the people in that district agreed. DeWitt Clinton of New York led his state in the creation of tax-supported schools and teacher training institutions.

Horace Mann, through his writings and work as secretary of the Massachusetts Board of Education, was a leader of the public school movement of the first half of the nineteenth century. Mann and other reformers called for longer school terms, consolidation of schools, the professional education of teachers, and many other improvements in the public school system.

The beginnings of the graded elementary school can be traced to 1818 with the founding of the Boston Primary School, which was organized into six classes. In 1836 William Holmes McGuffey began publication of his readers, which contributed to the establishment of elementary schools. By 1850, 45 percent of American youth attended school, and half of the states had established school systems before the Civil War.

The high school movement coincided with the elementary school movement, but high schools developed at a slower pace. The first American high school was established in 1821 in Boston, for boys who did not plan to attend college. In 1827 Massachusetts passed a law requiring towns of 4,000 or more to establish a high school, although some towns did not comply. By the outbreak of the Civil War, 300 high schools had been established in the United States, with one-third in Massachusetts (Pulliam and Van Patten 2007, 141).

Growth in the number of colleges was almost exclusively among private and denominational schools before 1860; the number of colleges in America reached 182 by the onset of the Civil War (Pulliam and Van Patten 2007, 143). Some state universities were also founded before the Civil War as part of the western movement. Pulliam and Van Patten (2007, 143) note that by 1850, pressure was mounting in the United States to establish scientific, agricultural, and engineering colleges.

In 1862 Congress passed the Morrill Act, which granted each state 30,000 acres of public land for each representative and senator in Congress. The land was to be used to support in each state at least one college that would teach agriculture, mechanical arts, and military science. While some states used the money to develop existing colleges, other states created new technical colleges. States controlled the administration and curriculum of these institutions.

After the Civil War, this country implemented the concept that public education should be free to all, in contrast to the European system in which elementary education was for the lower classes and secondary education was reserved for the elite. It was believed that a person could advance through schooling and climb the social ladder as far as his or her ability would permit. While education was perceived as important for making a living, education was expected to teach citizenship, morality, and self-improvement. The public education movement was enhanced in 1867 when President Andrew Johnson signed into law the creation of a national department of education.

The foundation of the American educational system was laid during the 18th and 19th centuries, and the role of education as a fundamental right was established by 1900. Although the funding of public education is still a topic of political dispute in the 21st century, our history clearly indicates that the American educational system throughout this country's history has diffused the values that were articulated in the U.S. Constitution. The public education system in the United States continues to evolve, and it is a fundamental and vital force in the information infrastructure.

How Education Contributes to Diffusion Theory

This country's schools and colleges fulfill the role of diffusing the knowledge necessary for youth to lead productive lives in our society. Schools teach fundamental skills, knowledge, and values that are the core of the American and global culture. In that role, teachers have developed theories of teaching, pedagogy, based on research that has developed theories of how people learn. This field of educational psychology provides us with theory about learning at various stages in the life cycle, addressing different learning styles and using various modes of communication. This area of knowledge and its implementation is an important part of the information infrastructure.

Erstad and Sefton-Green (2013) emphasize that schools are the dominant educational institutions in contemporary societies, but that learning occurs in a variety of formal and informal contexts, alone and in social groups, voluntarily and involuntarily, throughout our lives. Certainly library and information professionals are part of the "learning life" of adults and children whom they encounter.

Booth (2011, 36) outlines learning theory and effective instructional practices for library and information professionals to use in their work with clientele. Booth divides educational theory into three branches: learning theories that examine how knowledge is formed, instruction theories that apply teaching methods, and curriculum theories that focus on instructional content. We recommend this title as a well-written guide to understanding how people learn and how to structure an effective learning experience that is effective diffusion. Other recommended sources of learning, instruction, and assessment are Illeris (2009), Mayer (2011), Thomas and Brown (2011), and Erstad and Sefton-Green (2013).

These learning theories and teaching techniques can be applied by a teacher following a diagnosis of learning styles within a group as well as a diagnosis of the content that should be taught and appropriate teaching techniques. This diagnosis and concern for addressing learning styles is necessary for both traditional face-to-face instruction and for online instruction and various modes of distance learning. The traditional lecture mode can no longer be the only way of teaching; learners must be given opportunities to learn utilizing their preferred learning styles and be active participants, engaging with other learners and the instructor.

Computers and the Internet have reshaped the technology associated with the teaching/learning process. Sophisticated software has enabled the "virtual" library and classroom. Computers and the Internet provide gateways to learning that change our concepts of schools, instruction, teaching, and libraries. Stielow (2014) examines the revolutionary changes that confront education generally and libraries in particular. Although focused on academic libraries, Stielow provides insight into the cataclysmic changes that are influencing the roles of schools, teachers, libraries, and librarians.

Library and Information Professionals Promote Diffusion

In recent years, the role of library and information professionals has shifted from the more passive role of the past to a much more assertive role in providing library and information services to clientele. This shift, perhaps more appropriately termed a revolution, shifts the role of the library and information professional from that of dissemination to diffusion.

Special librarians led the way in this revolution, and we include school librarians in this category. Special librarians typically are the information specialists in an organization: a law firm or library, a medical school or other facility, a corporate enterprise, an information entrepreneur who contracts for information services, or a librarian in a public or private elementary or secondary school. These information professionals are the information experts surrounded by professionals with subject area expertise in such fields as medicine, law, educational theory, management, economics, and a variety of other professional fields. It has typically been their role to assess the information needs of their clientele and to help them utilize this information in their professional work. In doing so, they are engaged in the diffusion of information, often specialized information.

School librarians led the way, as reflected in their standards. In *Information Power: Guidelines for School Library Media Programs* (American Association of School Librarians and Association for Educational Communication and Technology 1988, 26), the role of the school librarian is defined as information specialist, teacher, and instructional consultant. These standards identified the role of librarians as providing the following:

- Access to information and ideas by assisting students and staff in identifying information resources and in interpreting and communicating intellectual content

- Formal and informal instruction in information skills, the production of materials, and the use of information and instructional technologies

- Recommendations for instructional planning to individual teachers as well as assistance in school-wide planning in curricular and instructional activities

These national standards, adopted jointly by the American Association of School Librarians (AASL) and the Association for Educational Communication and Technology, vaulted the school librarian into the teaching process as a partner with the classroom teacher.

The vital role of the school librarian as an educator is continued and enhanced in the most recent professional guidelines of AASL, *Empowering Learners: Guidelines for School Library Programs* (2009). This document recognizes the expanding role of technology in our lives and identifies the mission of the school librarian in helping students develop knowledge and skills for the 21st century:

> The mission of the school library program is to ensure that students and staff are effective users of ideas and information. The school librarian empowers students to be critical thinkers, enthusiastic readers, skillful researchers, and ethical users of information by:
>
> - collaborating with educators and students to design and teach engaging learning experiences that meet individual needs
>
> - instructing students and assisting educators in using, evaluating, and producing information and ideas through active use of a broad range of appropriate tools, resources, and information technologies
>
> - providing access to materials in all formats, including up-to-date, high-quality, varied literature to develop and strengthen a love of reading
>
> - providing students and staff with instruction and resources that reflect current information needs and anticipate changes in technology and education
>
> - providing leadership in the total education program and advocating for strong school library programs as essential to meeting local, state, and national education goals (8).

This emboldened role of the school librarian was accompanied by research and the development of models that facilitated this role of the school librarian as teacher and instructional consultant. Kuhlthau's (2003) research resulted in a six-stage Information Search Process model: (1) task initiation, (2) selection, (3) exploration, (4) formulation, (5) collection, and (6) presentation.

Kuhlthau's research further identified intervention levels by library and information specialists:

- Organizer: prepares the context for information use as a self-service facility

- Locator/lecturer: provides clients an overview of the information resources and services available, focusing on location of resources

- Identifier/instructor: provides instruction on specific resources and their use

- Advisor /tutor: provides numerous instructional sessions to help users understand various types of information resources

- Counselor: instructs over time how a user can identify and interpret information for specific purposes (Kuhlthau 2003)

One of the most widely used models is the Big 6, which identifies information problem-solving steps as (1) task definition, (2) information-seeking strategies, (3) location and access, (4) use of information, (5) synthesis, and (6) evaluation (Eisenberg and Berkowitz 1990). The authors found this model to be effective for library and information professionals working with information users in school, public, academic, and special libraries.

The Big 6 was used by the Kansas Association of School Librarians Research Committee to develop a model for teaching and assessing information skills of students. Named the Handy 5, this model identified five stages for teaching and assessing information literacy: (1) assignment, (2) plan of action, (3) doing the job, (4) product evaluation, and (5) process evaluation (Grover, Fox, and Lakin 2001). The model is intended to teach information skills with an emphasis on teaching the evaluation of information through the five steps. A second edition of this model provides additional examples and lesson plans for use in classrooms (Losey 2007).

The American Association of School Librarians has also published *Standards for the 21st-Century Learner*, guidelines for school librarians engaged in teaching information literacy skills within a school curriculum. This active engagement in the teaching process, as a partner with classroom teachers, is also found in the literature for college and university librarians. The Association for College and Research Libraries (2011) has guidelines for teaching information literacy. Also, ACRL has produced *Information Literacy Competency Standards for Higher Education* (2000), which are currently being revised. The professional literature of college and university librarians has adopted terms like "learning commons" and "research commons" to communicate the library as a center for providing students an opportunity to engage with library and information professionals, faculty, and other students in their pursuit of their learning.

The reader is advised to examine more closely the models described above. In addition, numerous resources for helping the information professional engage in teaching information literacy are now available. A sampling is represented in the "References" section below, including Burkhardt, MacDonald, and Rathemacher (2010); Cox and Lindsay (2008); Crane (2014); Daugherty and Russo (2007); Davis-Kahl and Hensley (2013); Eisenberg, Lowe, and Spitzer (2004); Farmer (2004); Heine and O'Connor (2014); Kuhlthau, Caspari, and Maniotes (2007); Ragains (2013); Smith (2011); Stielow (2014); and Thomas (2004).

The Role of Learning Libraries

Libraries have a central role in the transfer of information. Libraries have been viewed as repositories of information for the social good. The access and transfer of information is accomplished through the work of management and staff and through the establishment of policies. Knowledge transfer, on the other hand, occurs when knowledge is diffused from the individual to others. Diffusion works well when the potential user is highly motivated to learn, when finding information is relatively simple, and when the search process (access, cost, time, information pool) is reasonable. The library and information professional must be the leader and administrator who organizes the space, staff, and information resources to support teaching and learning. Examples of learning libraries are found in Chapter 9.

Summary

The purpose of libraries is the diffusion and utilization of knowledge. Libraries have shifted their focus from the identification, acquisition, storage, access, and preservation of information, a passive mode of services that accommodate dissemination, to a service-centered mode that supports diffusion of information and knowledge. Diffusion of knowledge is a key element in information transfer. Diffusion refers to the flow of knowledge from a source to the adopter. It denotes a process of social learning, embracing the transfer of a new idea within an individual or social system. Diffusion also implies a value shift.

The goal of information professionals is to create and operate information systems and services that accommodate the information needs and behavioral characteristics of a specific client population. This goal requires knowledge of human behavior associated with the acquisition and use of information. Information psychology is the field of information science concerned with the development of this theory of human behavior.

A professional possesses a particular knowledge that enables that individual to apply professional knowledge with a service as the product. The role of any professional, including library and information professionals, is that of diagnosing needs, prescribing a service that meets those needs, implementing that service, and evaluating the outcome of this interaction. Whether applying this diagnostic process to an individual or group, a critical component is the diagnosis of the individual's or group's needs, preferences, and cognitive styles. Information psychology helps us to understand the unique aspects of diagnosing the information needs of individuals.

At a societal level (including small groups, organizations, professions, occupations, and societies of various sizes), the social processes of creating, disseminating, diffusing, and utilizing information constitute "information transfer," and the diffusion process is central to the role of library and information professionals. Rogers (2003, 11) defines diffusion as the process by which (1) an innovation (2) is communicated through certain channels (3) over time (4) among members of a social system. Diffusion is a unique type of communication with a focus on the spread and understanding of a message or new idea.

A vital component of the information infrastructure is education, which is concerned with the diffusion of knowledge. Education is the diffusion of culture, and the purpose of a country's or organization's education system is to teach children and adults the core culture of that country or organization, the values associated with its history, and the necessary skills for a person to succeed in contemporary society.

In recent years the role of library and information professionals has shifted from the more passive role of the past to a much more assertive role in providing library and information services to clientele. This shift, perhaps more appropriately termed a revolution, shifts the role of the library and information professional from that of dissemination to diffusion, and the information professional is now actively engaged in the teaching of information literacy skills, assisting clientele in deriving meaning from information resources.

References

American Association of School Librarians. 2009. *Empowering Learners: Guidelines for School Library Media Programs.* Chicago: American Association of School Librarians.

American Association of School Librarians. *Standards for the 21st-Century Learner.* http://www.ala.org/aasl/standards-guidelines/learning-standards.

American Association of School Librarians and Association for Educational Communication and Technology. 1988. *Information Power: Guidelines for School Library Media Programs.* Chicago and Washington, DC: American Association of School Librarians and Association for Educational Communication and Technology.

Association of College and Research Libraries. 2000. "Information Literacy Competency Standards for Higher Education." http://www.ala.org/acrl/standards/ informationliteracycompetency.

Association of College and Research Libraries. 2011. "Guidelines for Instruction Programs in Academic Libraries." http://www.ala.org/acrl/standards/ guidelinesinstruction.

Baskerville, R. and J. Pries-Heje. 2001. "A Multiple-Theory Analysis of a Diffusion of Information Technology Case." *Information Systems Journal*, 11:179–180.

Booth, Char. 2011. *Reflective Teaching, Effective Learning: Instructional Literacy for Library Educators.* Chicago: American Library Association.

Burkhardt, Joanna M., Mary C. MacDonald, and Andree J. Rathemacher. 2010. *Teaching Information Literacy: 50 Standards-Based Exercises for College Students.* Chicago: American Library Association.

Cox, Christopher N. and Elizabeth Blakesley Lindsay. 2008. *Information Literacy Instruction Handbook.* Chicago: Association of College and Research Libraries.

Crane, Beverley E. 2014. *How to Teach: A Practical Guide for Librarians.* Lanham, MD: Rowman & Littlefield.

Daugherty, Alice and Michael F. Russo. 2007. *Information Literacy Programs in the Digital Age: Educating College and University Students Online.* Chicago: Association of College and Research Libraries.

Davis-Kahl, Stephanie and Merinda Kaye Hensley. 2013. *Common Ground at the Nexus of Information Literacy and Scholarly Communication.* Chicago: Association of College and Research Libraries.

Eisenberg, Michael B. and Robert E. Berkowitz. 1990. *Information Problem-Solving: The Big Six Skills Approach to Library and Information Skills Instruction.* Norwood, NJ: Ablex.

Eisenberg, Michael, Carrie A. Lowe, and Kathleen L. Spitzer. 2004. *Information Literacy: Essential Skills for the Information Age.* Westport, CT: Libraries Unlimited.

Erstad, Ola and Julian Sefton-Green, eds. 2013. *Identity, Community, and Learning Lives in the Digital Age.* Cambridge: Cambridge University Press.

Farmer, Lesley S. J. 2004. *Instructional Design for Librarians and Information Professionals.* New York: Neal-Schuman.

Greer, Roger C., Robert J. Grover, and Susan G. Fowler. 2013. *Introduction to the Library and Information Professions.* 2nd ed. Santa Barbara, CA: Libraries Unlimited.

Grover, Robert, Carol Fox, and Jacqueline McMahon Lakin, eds. 2001. *The Handy 5: Planning and Assessing Integrated Information Skills Instruction.* Lanham, MD: Scarecrow Press.

Heine, Carl and Dennis O'Connor. 2014. *Teaching Information Fluency: How to Teach Students to be Efficient, Ethical, and Critical Information Consumers.* Lanham, MD: Scarecrow Press.

Illeris, Knud, ed. 2009. *Contemporary Theories of Learning: Learning Theorists . . . In Their Own Words.* London and New York: Routledge.

Katz, E., M. L. Levin, and H. Hamilton. 1963. "Traditions of Research on the Diffusion of Innovation." *American Sociological Review*, 28:237–253.

Klein, S. S. and M. K. Gwaltney. 1991. "Charting the Educational Dissemination System." *Knowledge Creation, Diffusion, Utilization*, 12:246–248.

Kuhlthau, Carol Collier. 2003. *Seeking Meaning: A Process Approach to Library and Information Services.* Westport, CT: Libraries Unlimited.

Kuhlthau, Carol Collier, Ann K. Caspari, and Leslie K. Maniotes. 2007. *Guided Inquiry: Learning in the 21st Century.* Westport, CT: Libraries Unlimited.

Losey, Betsy, ed. 2007. *The Handy 5: Planning and Assessing Integrated Information Skills Instruction.* 2nd ed. Lanham, MD: Scarecrow Press.

Mayer, Richard E. 2011. *Applying the Science of Learning.* Boston: Pearson Education.

Pulliam, John D. and James J. Van Patten. 2007. *History of Education in America.* 9th ed. Upper Saddle River, NJ, and Columbus, OH: Pearson Prentice Hall.

Ragains, Patrick. 2013. *Information Literacy Instruction That Works: A Guide to Teaching by Discipline and Student Population.* Chicago: Neal-Schuman.

Reece, William J. 2005. *America's Public Schools: From the Common School to "No Child Left Behind."* Baltimore: The Johns Hopkins University Press.

Rogers, Everett. M. 2003. *Diffusion of Innovations.* 5th ed. New York: Free Press.

Smith, Felicia A. 2011. *Cybrarian Extraordinaire: Compelling Information Literacy Instruction.* Santa Barbara, CA: Libraries Unlimited.

Stielow, Frederick. 2014. *Reinventing the Library for Online Education.* Chicago: ALA Editions.

Thomas, Douglas and John Seely Brown. 2011. *A New Culture of Learning: Cultivating the Imagination for a World of Constant Change.* Lexington, KY: CreateSpace.

Thomas, Nancy Pickering. 2004. *Information Literacy and Information Skills Instruction: Applying Research to Practice in the School Library Media Center.* Westport, CT: Libraries Unlimited.

7

Putting Knowledge to Use

Chapter Overview

Diffusion, a vital stage in the information transfer process, precedes information and knowledge utilization. Utilization is the goal of information
transfer. In this chapter we explore the elements of utilization and the factors that can influence or hinder it. Utilization is not an all-or-nothing process, and library and information professionals must assess the individual
and social factors that may influence the utilization process. In addition to
exploring these factors, we provide a model that enables library and information professionals to design services that will enable utilization of new
knowledge in an organization.

Utilization in the Information Infrastructure

Utilization is a critical component of the information transfer process
and is enabled by diffusion, as explained in Chapter 6. When information is
created, the ultimate goal is utilization, and to understand the information
infrastructure, one must understand the elements of utilization.

The concepts of an information infrastructure and knowledge utilization have their foundation in much earlier times. The Greeks believed that
the production and use of knowledge benefits society. They understood that
a democratic government demands higher learning and higher education.
With the collapse of the Greek city-states, with the exception of the Roman
Republic, this attitude toward learning and the application of knowledge
was largely lost until its rediscovery in the Italian Renaissance.

It was not until the scientific revolution of the 16th and 17th centuries that discoveries by such luminaries as Copernicus, Galileo, Newton, and
Kepler, prompted the emergence of modern science in the fields of mathematics, physics, astronomy, biology, and chemistry and transformed the views of
society and nature, laying the foundation for the Age of Enlightenment.

In the 18th century, philosophers and intellectuals such as Diderot,
Montesquieu, Rousseau, and Voltaire, among others, created an unshakable
faith in human reason and the notion that continuous progress was possible. Their project, the encyclopedia, was to collect, organize, and distribute
all new knowledge that was based on reason—that is, secular, not religious,

knowledge. The assumption was that through scientific research and education, human change was possible. The result and orientation was toward utilization, fostering the Industrial Revolution during the 18th and 19th centuries.

Industrialization moved manufacturing from hand-production of products in homes and small shops to factories using powered machinery to mass-produce goods. Development of the steam engine played a central role in the Industrial Revolution, which also resulted in a revolution in transportation and communication, discussed in previous chapters of this book. The Industrial Revolution resulted in a revolution in the information infrastructure, including the way that new knowledge is put to use in society.

With the establishment of land-grant institutions following the U.S. Civil War, the intent was to make new knowledge available for use by the general public. Research from land-grant universities was instrumental in recovering from the Great Depression during the 1930s and 1940s.

Diffusion of innovation in agriculture was largely responsible for prompting American utilization studies. Rural sociologists employed in agricultural colleges and land-grant universities studied the spread of technology in farming. For example, innovation and utilization in the application of antibiotics to control disease, chemical weed sprays, pesticides, new seed varieties, and improved farm machinery were responsible for increased agricultural production. These innovations benefitted an increasing population and met the needs of government during World War II and the postwar demands for additional food requirements. The U.S. Department of Agriculture's Cooperative State Research, Education, and Extension Service is an example of a government policy initiative to increase diffusion and use of new research and products.

The post–World War II period showed an increase in government interest in knowledge utilization to stimulate economic growth through technology transfer. Space research and promotion of health, education and human services became policy objectives. A prominent example is President Johnson's Great Society and "war on poverty."

The computer and communication revolutions continued the shift toward a knowledge economy and a global economy. In a networked environment, knowledge is both an economic resource and a commodity. The world of Facebook, Google, LinkedIn, Twitter, and other social media highlights the increased importance of content and knowledge rather than only information dissemination. Customization of information services, creating packages of information and repackaging it, and information utility services add value to products that lead to greater utilization. As society created new knowledge ever more quickly and implemented new technologies that intensified the creation, mass production, dissemination, and diffusion of knowledge, the utilization process became more complex, and the information infrastructure's complexity has challenged utilization.

Utilization Defined

Larsen (1980, 421) raises the question, what is knowledge utilization? How do we define it, and is it situation-specific? Larsen notes that utilization studies are affected by "the world of political pressure, historical traditions, hunches and the like." Early assumptions were that an entire set of recommendations, as seen in the Iowa hybrid corn studies, must be accepted for utilization to take place. Nonutilization was not considered in that study.

However, like diffusion, utilization is a process with several stages: knowledge awareness is evaluated in a context of other options, and then utilization starts, perhaps with partial implementation, becomes part of the work routine and organizational policy, and eventually becomes integrated into the work process.

Stages of Utilization

A further amplification of classifying utilization and nonutilization is by Larsen and Werner (1981, 79–80). They note that implementation may occur in several stages:

- Complete implementation of information as presented;

- Adaptation of information—it is changed from its original presentation;

- Partial use of information; or

- Steps that have been taken toward implementation, although full implementation has not occurred.

Larsen and Werner (1981, 80) also classify stages of nonutilization:

- Information has been considered by a potential user but then is rejected;

- Nothing is done with the information; or

- Implementation of the information has not occurred, but is under consideration.

The above are attempts to understand how utilization potentially occurs in organizations. Clearly, utilization is not a forgone conclusion but instead depends on many variables and much individual and organizational effort and commitment.

Factors That Influence Utilization

The above explanations of utilization focus on the complexity of the processes of utilization and the many factors in play that can lead to either partial utilization or nonutilization. An individual's awareness, attitudes, past actions and behavior, and incentive are all factors in utilization of knowledge. While timeliness of information, objectivity, communication patterns, and political feasibility can play a role in utilization, a greater factor may be the need for an organization to survive in a globally competitive environment.

In this environment, knowledge utilization aims at increasing the employment of knowledge to solve problems, stay competitive, and improve the quality of life, thus pushing the desire for intervention and utilization of knowledge in an organization. These factors explain the complexity as well as richness of research in the field of utilization. According to Backer (1991, 225–240), it is a broad field encompassing areas such as technology transfer, dissemination and diffusion, research utilization, sociology of knowledge, organization change, and policy research. Most importantly, utilization involves designing strategies that help put knowledge to use.

Backer (1991, 233–234) identifies four critical principles that affect utilization:

1. Knowledge utilization requires individual and organizational *change,* which can be both mechanically difficult and psychologically threatening for those considering the implementation of some new program or procedure.

2. Knowledge utilization requires *resources*; money, materials, and personnel are needed for any significant change especially if the change takes place within a complex organizational or social environment.

3. Adopters of innovation must be convinced that the innovation *will work* in their particular setting, meeting specified needs over time without excessive side effects or unreasonable cost.

4. Innovation adopters must be *aware* of the program or practice—an awareness that seems self-evident, yet in many cases, worthwhile innovations do not get transferred simply because potential users do not know about the new program or practice.

In this study Backer also provides "strategies" for the above challenges to utilization. They are as follows:

1. Interpersonal contact. To get an innovation used in a new setting, there needs to be direct, personal contact between staff members of the potential adopting organization and those with direct knowledge about the innovation.

2. Planning and conceptual foresight. A carefully thought-out plan for how the innovation will be adopted in a new setting is essential to meet the challenges indicated earlier.

3. Outside consultation on the change process. Such consultation provides help in designing the change effort efficiently and offers some useful objectivity about what needs to be done.

4. User-oriented transformation of information. What is known about an innovation must be translated into language that potential users can understand readily, abbreviated so that attention spans are not exceeded, and made to concentrate on the key issues of "Does it work?" and "How can I replicate it in my organization?"

5. Individual and organizational championship. An innovation's chances for successful adoption are much greater if influential staff members and organizational leaders express enthusiasm for its adoption.

6. Potential user involvement. Everyone who will have to live with the results of an organizational change must be involved in planning for innovation adoption, both to get a range of suggestions for how to undertake the adoption effectively and to facilitate "felt ownership" of the new program (thus decreasing resistance to change).

7. Partnership, knowledge and understanding of the innovation, careful planning, champions for the impending change, and ownership are critical success factors.

In summary, utilization is not an all-or-nothing process; it is complex and requires a careful assessment of the environmental context and the personnel engaged in the utilization. Knowledge utilization is integral to change in an organization. Also, both individual and social variables influence utilization—the same factors that influence diffusion, as described in Chapter 6. Following are models that provide guidance for utilizing new knowledge in an organization.

Social and Environmental Issues That Influence Utilization

Researchers from a number of disciplines have proposed a variety of knowledge utilization models that explore the complex social and environmental issues surrounding knowledge utilization. Here we focus on the work of Carol H. Weiss.

The knowledge-driven model of Weiss (1977, 13) states that basic research discloses some opportunities that may have relevance for public policy. Applied research is conducted in order to define and test findings of basic research for practical action; if all goes well, appropriate technologies are developed to implement the findings, whereupon utilization occurs.

This model suggests a relatively simple, linear, orderly sequence of processes, not an interactive one. Weiss argues that this model approximates physical science more than the social sciences, but we question the linearity and approximation to the physical sciences. Our view is that models applied to the social sciences may appear to be linear, but the variables imposed by the environmental context result in nonlinearity.

However, Weiss's model is useful in library and information science applications. This problem-solving model refers to the direct use of social research for policy decision-making (Weiss 1977, 11–12). Research is instrumental in solving social or political problems. A problem exists, information is lacking, research provides missing knowledge, and a solution is reached.

This model is based on the assumption that policy-making is a rational process, and the model ends with policy choice. It does not concern itself with issues such as relevance, choice, adoption, implementation, and many other factors necessary for successful utilization.

In the "enlightenment model," Weiss (1979) acknowledges the general and broad impact of knowledge on the policy-making process. She notes that research rarely provides a solution to solve a specific policy problem. Instead, research background data, empirical generalizations, and ideas are supplied to shape policy-makers' thinking. Research results shape their conceptualization of a problem's solution. The "enlightenment" is not ideal, since research results come to the attention of policy-makers conceptually and not necessarily at a time of need. The process can result in "endarkenment as well as enlightenment" (Weiss 1979, 430).

The Weiss (1977) "social interaction model" views society as a network of roles and channels of communication, both forming barriers and overlapping connections. It is a nonlinear process, not an orderly moving from research to decision, but instead a set of interconnections and feedback. It is closer to the system perspective that is highlighted as a fundamental conceptual framework for modeling information transfer as described in Chapter 4.

The models described above reinforce the importance of environmental issues that have an impact on utilization. The reader may wish to review the section "The Environmental Context of Society" in Chapter 2. These variables influence us as individuals and groups as well as all of the information transfer processes: creation, production, dissemination, diffusion, organization, and utilization of information and knowledge.

The environmental and social context includes all of the following: culture, physical geography, political structure, legislation and regulations, the economic system, technology, and information policy. Legislation, regulations, and information policy can help or hinder effective utilization. For example, copyright, which safeguards the rights of authors and creators of information and knowledge, must be considered when implementing a copyrighted document so that the rights of the author are safeguarded. Furthermore, regulations by a legislative body, organization, or private enterprise may block or shape the utilization of knowledge. For example, nonpartisan organizations like the League of Women Voters might refuse to diffuse or otherwise utilize information disseminated by a political party.

Another factor in utilization is the use of technology. It's safe to say that all of us have attended presentations that were hindered by faulty technology or lack of knowhow by the presenter. One of the authors recently was deprived of phone and Internet service because his provider, an international telecommunications company, suffered a technical malfunction, and his "bundle" of services was short-circuited for a day.

All of the environmental factors together account for the fragility of diffusion and utilization. Technology especially is a vital but fragile component of the global information infrastructure, and malfunctions are common and must be anticipated. Cell phones, Internet connections, television signals, and networks are subject to technical malfunctions and weather conditions. A strong wind or violent thunderstorm can interrupt satellite transmission for minutes and perhaps hours. Although technology is a boon to diffusion and utilization, the wise information professional plans for a variety of delivery modes when relying on technology.

Nevertheless, technology provides the opportunity for the information professional to diagnose a client's information needs and to apply appropriate technology to address those needs. It is essential that an information professional develops a close working relationship with clients, to understand their information needs and level of technology skill so that an information package can be customized for the client's use.

Individual Variables That Influence Utilization

As discussed in Chapter 6, diffusion must be thought of in two ways—it occurs in individuals and in groups. Similarly, we must think of utilization as both an individual and a group (or social) process. The same environmental variables that influence utilization of knowledge in groups (see above) also influence individuals.

An interesting and appropriate development in the study of knowledge utilization is the influence of constructivist learning theory. This theory emphasizes that knowledge is not a thing, a static object to be transmitted and passively received. Constructivism assumes that new knowledge is about sense-making; knowledge is filtered by a learner's preexisting experience and environment. Knowledge is fluid and negotiated between the creator and the user—teacher and student. Users assume control over learning and are actively engaged in problem-solving and constructing their own understanding. Thus, utilization models benefit from the diffusion models and the role of "adoption" in the diffusion process.

From a systems perspective, utilization is a part of the dynamic information transfer process. Conceptually, utilization begins with the knowledge creation process, a partnership of the researcher and potential user. The goal of creation, dissemination, production, organization, diffusion, and preservation of information should be its utilization. Thus, information need, context, prior experience, values and beliefs, channels of communication, media formats, and language used need to fit a target audience.

Knowledge products must be understood and must accommodate the user's existing knowledge to be diffused and used. Utilization is ultimately a process of involving the potential user in adopting new knowledge and accepting change. The implications of this perspective for librarians and information professional in the design and delivery of services are tremendous, as discussed in the following section.

Utilizing Information and Knowledge in an Organization

Information and knowledge can be categorized in order to plan for utilization in an organization. The authors, in their professional work, teaching, and writing (Greer, Grover, and Fowler 2013, 139–147) have defined the following categories of information and knowledge use:

- Teach or learn: The educational function of information and knowledge

- Enjoy: The recreational function

- Appreciate: The cultural function

- Decide: The informational function

- Create new information or knowledge: The research function

- Find or locate: The bibliographic (or identification) function

These uses of information and knowledge are the basis for the services provided by library and information agencies—services with the goal of information diffusion and knowledge utilization. We will discuss each service or function in more depth below.

Educational Function

The educational function of libraries and information agencies supports learning—the diffusion of information. This function is one of the oldest forms of service provided by the profession.

In recent years, librarians and other information professionals across the profession have become more engaged in the teaching/learning process. School and college/university librarians have led the way, collaborating with faculty to teach students information skills so that they can be effective users of information and knowledge. "Information literacy" is the term often applied to the ability to use information effectively. However, teaching effective use of information in this knowledge age is an important function of medical, legal, public, corporate, government, and all types of special library and information service agencies.

Developments in telecommunication and instructional technology have extended the classroom from an enclosed space to a virtual classroom without boundaries. Distance learning enables schools, colleges, and other organizations to provide instruction to students anywhere in the world via the Internet with two-way audio and video communication. Software designed for distance learning enables students to interact in real time or asynchronously and to submit assignments to the instructor electronically. Regardless of the agency offering the instruction, and regardless of the instructional format, the information professional can assist by providing resources for both the instructor and the learners.

Cultural Function

The cultural function of libraries and other information agencies provides clientele the best examples of a society's cultural heritage as represented in literature, history, music, dance, and various art forms. In corporate or special libraries, the culture may be that of a profession (medical, law) or an organization (IBM).

The challenge of library and information professionals is to select the "best" available information resources representative of the culture. In a corporate or institutional setting, the clientele may be instrumental in determining which are the best journals, databases, Web sites, blogs, and books to procure for the collection. Law librarians consult their primary clientele of attorneys and other legal practitioners regarding the electronic and print books and journals that are considered the most reputable and useful, and they weigh that input with their professional assessment of new resources that are challenging the traditional resources in terms of content and price.

The cultural function includes the sharing of cultures of various ethnic groups—the stories, music, videos, motion pictures, literature, and customs of immigrants and of other international cultures. Increasingly we are citizens of a global society, and libraries are repositories for representative artifacts of the cultures of our world.

The cultural function includes a variety of formats: books, periodicals, audio recordings of various formats, Web sites, and video recordings. Consequently, information professionals must be aware of the technologies, current and emerging, that record various cultural products.

Recreational Function

The recreational function is synonymous with "enjoyment" and is concerned with the use of information resources for leisure. A variety of formats are made available for recreation, including novels, poetry, periodicals, and nonfiction books, as well as audio and video recordings and games.

The rationale often given is that the recreational function is a way of reaching people who may not otherwise use the library, or who cannot afford to buy the resources, or simply don't want to pay for them if they are "only" recreational. The librarians may steer clientele, once they are in the library, to other resources and services that will benefit them. Our point is that recreational use of information and knowledge resources is still a viable type of utilization.

The Internet has transformed all information functions, including the recreational function. Information users access the Internet for games, music, movies, and other sources of entertainment that can also be educational, leading to the term "edutainment." The Internet is a major source for the recreational function, a source that undoubtedly will become more prominent in years to come.

Research Function

The research function results in the creation of new social knowledge. The use of "research" here is that systematic, formal process that requires (1) a literature review to determine the state of the art in an area of inquiry, (2) a careful articulation of research questions or hypotheses, (3) a systematic plan and effort to collect data, (4) analysis of the data, and (5) a report of the results, including generalizations extrapolated from the data.

This research may be the formal, carefully articulated and conducted study that conforms to the scientific paradigm of a discipline or field, or it may be the problem-oriented, practical research conducted by members of a profession. The latter research, intended to solve professional problems, is often called "action research." Both types of research contribute new knowledge to a discipline or field of study.

Engagement in the research process depends on the environment of the information agency. School librarians are rarely engaged in the research function, except in those rare cases when a teacher or principal may request help as they conduct "action research" to solve a professional problem or to fulfill an assignment in a graduate class. However, school librarians and other information professionals may themselves engage in action research to address professional problems and publish the results.

In academic libraries, librarians are on call and can be consulted by faculty to assist with research through in-depth consultation. Information professionals in research libraries and archives often engage with researchers for a long period of time. For example, an archivist often works with researchers by appointment, and a librarian might work several hours with a researcher on one question. Regardless of the type of research or the environment where it is conducted, use of information and knowledge as part of the research process is an essential part of the information infrastructure.

Informational Function

The information function provides information for decision-making—utilization. Providing information to people instead of teaching them *how to use* information sources is not a new concept in the library and information profession; however, implementation of information service has been slow to evolve.

The information function is often called "reference service" or "information service" in today's libraries and information centers. Information service is found in nearly all libraries and includes the teaching of information skills (really the educational function), "ready reference," and regular reference service. "Ready reference" is the term used for questions that can be answered very quickly by consulting one or two information sources.

In most information agencies, some type of information service is provided, sometimes for a fee. If a service requires an extraordinary amount of a professional's time, the client may be charged—especially in special libraries. Even some public libraries have charged fees for their services when that service was beyond the "standard" service for clientele—what we call "reactive" level service, described below.

A recent example of proactive information service is the concept of the "embedded librarian." Some public, academic, school, and special libraries send library and information professionals into their communities to attend faculty meetings or city commission meetings, to visit with department heads, and to interact in order to discern information needs, serving as liaisons between the agency and the library. Then the information professionals work with other members of the library staff to plan and deliver information services to address those needs.

These services sometimes cross functions, according to our definitions, but they are launched to meet specific needs of individuals. As technology becomes more sophisticated, especially with the integration of artificial intelligence, the ability for computers to diagnose the information needs of clientele will be a major contributor to the advance of information services in the future.

Bibliographic Function

An important information function is that of organizing information for storage, retrieval, and use. In library and information agencies, the organization of information for client use is accomplished through bibliographies, indexes, pathfinders, and other finding aids.

Once again, to implement the bibliographic function effectively, the information professional must know the needs of clientele. Then a finding aid can be developed to address those needs.

A distinct variation of the bibliographic work of most librarians is the bibliographic function of the archivist. A finding aid prepared by an archivist is equivalent to the title page of a book. Unlike the cataloging information used by nearly all librarians today to catalog books and other formats, such descriptive cataloging information is not provided to an archivist, who must study a collection of items (for example, manuscripts, photographs, realia), understand the collection and its importance, organize it, and then develop a finding aid. Standards for describing archives have been developed by the Society of American Archivists and include the following: creator information, an abstract of the collection, an explanation of the organizational scheme, and inclusive content notes—a succinct bibliographic history.

As technology has advanced, the bibliographic function has been transformed. For example, the "open access" movement is a new perspective on the bibliographic function. Academic librarians are working with researchers to organize and store data and research reports in institutional repositories that may be open to the public. This application of the bibliographic function teams an information professional with researchers to help them store and retrieve their research data, further engaging the information professional in the research function as researchers complete their data collection, analysis, and report writing.

The bibliographic function has been, and continues to be, a vital aspect of the information professional's work and of the information infrastructure. As the amount of information and knowledge proliferates, the information professional's ability to organize data and information and to provide finding aids becomes more valuable.

Levels of Service

The above services, which attempt to promote knowledge utilization, may be offered with differing outcome goals. A model for outlining levels of service was introduced by Greer and Hale (1982) and uses the following terms:

Passive

"The passive level of service consists of a process of choosing, acquiring, and organizing materials on the library shelves for the user to discover" (Greer and Hale 1982, 359). The information professional does not attempt to assist with the understanding or use of the information. For example, an information center provides a collection of books or journals, or it creates a database for use by clientele with no assistance.

Reactive

An example of the reactive level of service, in which professional assistance is provided on request of the clientele, is the traditional "reference service" offered in most libraries. Another example is the provision of finding aids (bibliographies. mediagraphies, or webographies) upon request. "The bulk of the material chosen and programs offered is based on the librarian's best judgment of relevance. Input from the community is welcomed when it arrives, but it is not collected systematically" (Greer and Hale 1982, 359).

Assertive

This level of service addresses the known information needs of clientele as identified by an information needs analysis. The assertive level of service anticipates information need and adds value to the information in order to enhance diffusion and utilization. A close working relationship between information professionals and clientele is essential for this level of service.

An example is the delivery of information to clientele without their asking; a university librarian e-mails a faculty member a newly arrived journal article for a literature review the professor is preparing. A school library media specialist, knowing the calendar of subjects to be taught to seventh-grade history students, arranges a meeting with history teachers to plan a history unit that integrates the teaching of note-taking skills.

The keys to an assertive level of service are (1) systematic collection of community data, from which needs and interests are inferred; (2) development of a customized collection; and (3) dissemination of information to meet the inferred and expressed needs. The central focus of this level of service is the community and its people (Greer and Hale 1982, 359–360).

These levels of service can be applied to the various information functions or services described earlier in this chapter. Those services can be provided at different levels, as depicted in Table 7.1.

The Table 7.1 matrix is a model for considering the variety of services that can be offered in a library or information agency. The analysis of need and the development of services in response are the essence of the library and information professions and provide valuable services in the information infrastructure that result in information and knowledge utilization.

Table 7.1   Information Services and Levels of Service

Function	Passive Level	Reactive Level	Assertive Level
	A collection of resources is provided.	Services are provided when requested by clientele.	Based on a needs assessment, client needs are anticipated and value-added service is provided.
Educational			
Cultural			
Research			
Recreational			
Informational			
Bibliographic			

Summary

Diffusion of information must precede information and knowledge utilization. In this chapter we have explored the steps of utilization and the factors that can influence or hinder utilization. As society created new knowledge ever more quickly and implemented new technologies that intensified the creation, mass-production, dissemination, and diffusion of information and knowledge, the utilization process became more complex, and the information infrastructure's complexity has caused challenges to utilization. Effective utilization of information requires collaboration between information professional and client—a partnership that enables the professional to customize information using appropriate technology.

Utilization is not an all-or-nothing process, and library and information professionals must assess the individual and social factors that may influence utilization. However, like diffusion, utilization is a process with several stages: knowledge awareness is evaluated in a context of other options, and utilization starts perhaps with partial implementation, becomes part of the work routine and organizational policy, and eventually becomes integrated into the work process.

Utilization is complex and requires a careful assessment of the environmental context and the personnel engaged in the utilization. Knowledge utilization is integral to change in an organization. Also, both individual and social variables influence utilization, the same factors that influence diffusion.

Environmental issues have an impact on utilization. Culture, physical geography, political structure, legislation and regulations, the economic system, technology, and information policy are variables that influence us as individuals and as groups within society, and these variables influence all of the information transfer processes: creation, production, dissemination, diffusion, organization, and utilization of information and knowledge.

Library and information professionals promote utilization of information and knowledge by designing services around the functions of information and knowledge: the educational function, recreational function, cultural function, informational function, research function, and bibliographic (or identification) function. These uses of information and knowledge are the basis for the services provided by library and information agencies, services with the goal of information diffusion and knowledge utilization.

The analysis of need and the development of information services are the essence of the library and information professions and provide valuable services to promote information and knowledge utilization, essential components of the information infrastructure.

References

Backer, T. E. 1991. "Knowledge Utilization: Third Wave." *Knowledge: Creation, Diffusion, Utilization*, 12:225–240.

Greer, Roger C., and Martha L. Hale. 1982. "The Community Analysis Process." In *Public Librarianship: A Reader,* ed. Jane Robbins-Carter, 358–366. Littleton, CO: Libraries Unlimited.

Greer, Roger C., Robert J. Grover, and Susan G. Fowler. 2013. *Introduction to the Library and Information Professions.* 2nd ed. Santa Barbara, CA: Libraries Unlimited.

Larsen, Judith K. 1980. "Knowledge Utilization: What Is It?" *Science Communication* 1:421–442

Larsen, Judith K., and Paul D. Werner. 1981. "Measuring Utilization of Mental Health Program Consultations." In *Utilizing Evaluation: Concepts and Measurement Techniques,* ed. James A. Ciarlo, 77–96. Beverly Hills: Sage Publications.

Weiss, Carol H. 1979. "The Many Meanings of Research Utilization." *Public Administration Review* 39:426–431.

Weiss, Carol H., ed. 1977. *Using Social Research in Public Policymaking.* Lexington, MA: Lexington Books.

8

Navigating the Information Infrastructure

Chapter Overview

The purpose of this book is to unravel the complexity of the information infrastructure to help librarians and other information professionals in their role as navigators. In this chapter we explain how our model of the information infrastructure can be a guide for navigating the vast information infrastructure and for helping clientele to evaluate information, to find meaning, and to use information effectively. We examine different functions of information and provide examples of an information professional's role in this dynamic digital environment in which we live.

Overview of the Information Infrastructure

Information professionals are knowledgeable of the information transfer processes, and their role is to diagnose information needs of individuals and groups and to prescribe information sources and services to address those needs. In the diagnostic process, the information professional must be able to assess the information resources at each of the stages of the information transfer process and to assist the information consumer to select the appropriate information for her purposes. In so doing, the information professional is helping the information consumer to make sense of the information in order to utilize it; the information professional is facilitating the diffusion and utilization of information. To enable information professionals to be successful guides to the information infrastructure, we will use the information infrastructure guide that is outlined here in Table 8.1.

In this chapter, we will explain this table as a way of simplifying the information infrastructure and the evaluation of information at each stage of the information transfer process. We'll begin at the beginning with the creation and recording of information.

Table 8.1 Information Infrastructure Guide

Information Transfer Processes	Organizations	Professions	Research	Popular Information Sources
Creation/recording	• Colleges & universities • Think tanks • Professional organizations • News organizations • Radio & TV • Corporate researchers • Government agencies	• Professors • Scientists • Researchers • Journalists • Library & information professionals • Webmasters • Writers • Information entrepreneurs	• Who is/are the researcher(s)? • What are their credentials? • Is the sponsoring organization reputable? • Are the research questions appropriate? • Are the methodologies appropriate? • Are the results credible? • Does the work contribute to the field?	• Who is reporting? • What are their credentials? • Is the sponsoring organization reputable? • Is the reporting based on direct observations or interviews? • Is the reporting current? • Are multiple perspectives apparent? • Are the details accurate? • Does the report add to known information on the issue?
Mass-Production	• University repositories • Book publishers • Research journals • Newspapers • TV and radio stations • Web sites • Blogs • Media producers	• Researchers • Librarians & information professionals • Editors • Publishers • TV & radio producers • Webmasters- • Electronic publishers • Information entrepreneurs	• Is the research report peer-reviewed? • Is the publisher reputable?	• Is the publisher or production company reputable? • Is the content peer-reviewed?
Dissemination	• Professional organizations • Libraries & information agencies • Bookstores • Government bookstores and Web sites • TV and radio stations • Web sites • Blogs • Social media	• Administrators • Librarians & information professionals • Reviewers • Bookstore staff • Webmasters • TV & radio producers & on-air staff	• What agency is responsible for the distribution? • If a Web site or social media, is the purveyor acknowledged as reputable? • Is the research reviewed by an authentic source?	• What entity is responsible for the distribution? • If a Web site or social media, is the purveyor acknowledged as reputable? • Has the report been reviewed externally by a reliable source? • What is the purpose of the report or document? The intended audience? • Is the content appropriate for the intended audience?

Stage	Organizations	Professionals	Questions	Questions
Information Organization	• Libraries & information systems • Bibliography publishers in all formats • Webographies • Professional organizations • Publishing companies • Government agencies or nonprofit agencies	• Librarians & information professionals • Publishers • Editors • Webmasters • Indexers • Subject specialists	• Is the document found in WorldCat or other bibliographies? • Is it accessible through a general index? • Is the item representative of or "the best" of its type? • Is the document accessible through a specialized subject index or bibliography?	• Is the document accessible through an index or bibliography? • Can previous reports and other documents be retrieved easily? • Is the item of a quality worthy of its cost? • Is the document accessible through a specialized subject index or bibliography?
Diffusion	• Schools • Colleges and universities • Churches • Private enterprise • Libraries • Social organizations • Blogs	• Teachers & other academic staff • Clergy • Managers • Librarians & information professionals • Sales & marketing professionals	• Is the content included in university or school curricula? • Is the content appropriate for the intended audience? • Has the content been reviewed by an acknowledged source? • Is the content cited by peers?	• Is the report appropriate for the intended audience? • Is the format appropriate for the intended purpose and audience? • Is the content summarized and discussed or recommended by respected professionals?
Utilization	• All organizations	• All professionals	• Is the content utilized by individuals, professionals, or organizations? • Is the content cited by others in the field?	• Is the content utilized by individuals, professionals, or organizations? • Is the content cited by others?
Preservation	• Libraries • Archives • Museums • Public and private agencies • Think tanks & nonprofit agencies	• Librarians & information professionals • Archivists • Curators • Office managers • Bloggers	• Is the content pertinent to the organization's mission? • Is the format usable?	• Is the content pertinent to the organization's mission? • Is the format usable?
Discarding	• Libraries • Archives • Museums • Public and private agencies • Think tanks & nonprofit agencies	• Librarians & information professionals • Archivists • Curators • Office managers • Bloggers	• Is the content outdated or superseded? • Is the format obsolete, flawed, or otherwise unusable?	• Is the content outdated or superseded? • Is the format obsolete, flawed, or otherwise unusable?

Creation/Recording

One of the first questions to ask when evaluating information is, "What is the authority for this information?" In other words, who is responsible for creating this information? Is it an individual? A team? What organization sponsors the research or creation of this information? What are the credentials of this person or organization? What are the qualifications that suggest that the information produced has been produced using sanctioned methodologies and based on a systematic review of previous research in the field?

If this is research, does the researcher have a doctorate in the field? What is the sponsoring institution—a university, think tank, or other organization? If it is a reputable research university, one can be reasonably certain that the research is solid, because faculty members are submitted to careful scrutiny when they are hired, and they are evaluated each year until tenure is granted, usually following the sixth year of service. The department or school has ascertained that the researcher is qualified to conduct the research and that his work reflects the dominant paradigm in the field.

If the sponsoring institution is not a university, the researcher(s) may reflect the paradigm that the organization is promoting. The organization's Web site, annual reports, and other publications can be studied to determine the purpose of the organization; its values will be stated or apparent upon examination of these documents.

Another question to address is the quality of the research. Are the research questions appropriate for the study? Are the methodologies appropriate and properly applied? Is the analysis logical, and are the findings feasible? Are the findings meaningful? Are they a significant contribution to the field?

Authority for more informal information sources can be more difficult to determine. Again, the important question to address is this: Who is responsible for this information? Is it a professional organization? Who is the author, reporter, or other person responsible for creating the information? What are her qualifications? What are the sources of the information reported—direct observations, interviews, or other sources? Is the information current and accurate? Is the information unbiased? Does the information add to the known information on this issue? If the new information conflicts with other known sources, is the difference explained with appropriate evidence to support it?

A difficult issue to address is the perspective of the reporter or author—is he or she expressing a point of view or bias? If so, what is it? Perhaps the organization that the reporter represents is a clue. Newspapers or other news organizations develop a reputation for a bias or for more objective reporting. In recent years news organizations have freely expressed their viewpoint, some engaging in polemics. For example MSNBC makes no secret that it presents progressive views, while Fox News is more conservative in its political orientation. Is it possible to determine where the organization or the information lies on the political spectrum?

An effective means of assessing the authority is to determine the affiliation of the author or reporter of the information. However, as noted in our discussion of the creation and recording process, individuals can easily enter the information arena by writing and mass-producing their ideas via social media, blogs, and self-publishing. In such cases, the discerning reader should look for the author's credentials for writing, recognizing that a disreputable author can fabricate his qualifications as well as his information.

Mass Production

As noted in Chapter 5, the creation, recording, and mass-production of information have merged in the digital age. Now recording of information can be instantaneous with reproduction when a news reporter sends a message via social media or a television reporter describes a news event in progress on air. Storing a research report in a university repository that provides open access to that report also is a merging of recording and reproduction. The lines between the two information transfer processes, creation/recording and mass production, are blurred.

Organizations engaged in mass production of information continue to be publishers of books and journals, newspapers, television and radio stations, Web sites, blogs, and producers of various types of media: educational and entertainment motion pictures, audio and music recordings, photography, etc. Usually an organization engaged in mass production uses a variety of media, including traditional printed resources and electronic resources.

In the evaluation of mass-produced information, authority continues to be a major consideration; is the information peer-reviewed or at least reviewed internally within the organization during the production process? Is a report of a research project reviewed by a blind review of peers in that discipline or profession before it is accepted for publication? Does an editor or editorial board review a newspaper article or recorded television interview before it is aired? Does the organization have a reputation for accurate, unbiased reporting of research or news events?

If the persons responsible for the information are solely responsible for the content of the blog post, Web site, tweet, or article, the information consumer should review the information with a critical eye. In this digital age, when nearly anyone can enter the public dialogue via social media, careful scrutiny must be given to information that is available for public consumption. Information professionals can provide important evaluation services for the public.

Dissemination

The dissemination process has been decentralized and melded with mass production. Dissemination includes the earlier methods of printing newspapers, books, journals, and other documents on paper, but those methods have been supplemented with digital documents. This merging of processes makes the evaluation of resources more complex. Once again, the responsible organization is a key to the authority of the information provided. Professional organizations, libraries, bookstores, government bookstores and agencies, Web sites, television and radio stations, and blogs are engaged in dissemination. Because individuals with technical skills can disseminate information worldwide through the Internet, the dissemination source is an important consideration.

If the disseminator is an individual, that person's credentials should be examined to determine the authority of the information. Is the individual responsible for the information a knowledgeable professional with knowledge of the content area? Is she a system administrator, librarian, reviewer, webmaster, producer, or experienced staff member?

Information Organization

The library and information professions have a history as bibliographic control experts. Most libraries and information agencies of all types now have online catalogs or bibliographies that provide access to their collections by author, title, subject, and keywords. As a result, these catalogs are available on phones, personal computers, and tablet computers for easy access nearly anywhere. Is the document under consideration part of the information infrastructure that is subject to the general organization schemes of the library profession? Is the document found in WorldCat, the worldwide catalog that also indicates libraries that house specific titles? If the document is a periodical article, is it indexed in a reputable index or bibliography? Is the document in question of a quality that justifies its cost for purchase (if applicable), storage, and retrieval in the information system?

In organizations of all sizes, the organization, storage, and retrieval of ephemeral information like e-mail communications may not be systematic. If systematically stored, retrieval by keywords can be accomplished quite easily. Does the organization under consideration have such a systematic storage and retrieval plan?

While libraries and the library profession attempt to organize all information using a general classification scheme, professions and disciplines may organize books, research reports, and journal articles using terms peculiar to their profession or discipline. Librarians and indexers with subject-area expertise produce indexes, abstracts, and catalogs in their particular fields. An evaluation of specific documents should address the following questions: Is the document accessible through a reputable specialized index or catalog? Is the index or catalog the product of a service aligned with the professional or discipline field? Is a similar information scheme used for more ephemeral documents and communications?

Diffusion

Diffusion of information is the critical function whereby people make sense of information so that it can be utilized. Educational institutions and organizations have many communication channels to utilize in their educational missions. Private enterprise, religions, and organizations of all types are engaged in teaching. Consequently, the consumer has many choices in selecting educational opportunities in the marketplace. Likewise, those individuals and organizations offering educational experiences have many options for providing the programs they wish to offer. Bloggers can promote diffusion through their interpretation and discussion of issues.

The proliferation of information available for instruction requires an information specialist to help instructors use the Internet and other resources for teaching. This information specialist can help the student to identify resources to help in the learning process and the completion of learning assignments.

In the evaluation of learning resources, the authority of the resources is a first consideration (see the "Creation/Recording" and "Mass Production" sections above). Is the university, private enterprise, church, government agency, or other organization a reputable source for this information? If the author is an individual, does that person have the appropriate education and experience to claim expertise in the subject matter? Is the content appropriate in reading level and subject area for the intended audience? Has the resource been reviewed favorably by a reliable reviewing source? See Chapter 6 for a more complete discussion of diffusion.

Utilization

Consumers of information must be able to take information that has been disseminated and diffused through libraries, schools, or other channels of communication and use that information in their lives. It is the role of information professionals to help people understand information, to give it meaning, and perhaps to assist in interpretation so that people can use information in their personal or professional lives.

Utilization can be measured by observing the information consumers or by interviewing them. In a formal or research context, utilization of research can be measured by the citations one receives from other researchers and scholars. Popular resources likewise can have their utilization documented by citations and references in the popular media and social media. See Chapter 7 for an in-depth discussion of utilization of information.

Preservation

The ephemeral nature of social media and other digital documents is a factor in the preservation process, and individuals must make decisions regarding the value of their communications. Digital communications for organizations can be important for decision-making and policy-making; consequently, websites, digital documents, and e-mail should be included in the preservation policies of an organization.

As hardware changes are made within an organization, capability for retrieving digital documents is a consideration in preservation policy. Another consideration is the protection of digital documents by systematic storage and organization so that the documents are not lost.

As with utilization, a primary question in preservation/discarding decisions is the utilization of the document. Is the content of value to the organization now and in the future? Does the information address the mission of the organization? Regardless of the format (paper or digital), is it usable by the potential users?

Discarding

The process of discarding digital records is much the same as with paper documents. Items may be discarded if the content is irrelevant, obsolete, outdated, or superseded. If the digital document is flawed or cannot be retrieved because of obsolescence, it should be discarded.

Resources in all formats should be evaluated regularly. When resources are no longer contributing to the mission of the agency, they should be discarded. Policies should be developed to provide for a regular and systematic review of documents held by an organization.

Summary

The proliferation of information through a wide variety of both traditional and newer media is confusing, and the information consumer can feel overwhelmed. By applying the information transfer model for evaluating information at each stage of the information transfer process, the information consumer can be a more intelligent user of information. The information infrastructure matrix and guide in this chapter can be helpful to both information professionals and consumers to assist in the effective diffusion and utilization of information.

9

The Library and Information Professions in the Information Infrastructure

Chapter Overview

This chapter examines the role of today's library and information professionals as they work within the emerging information infrastructure. In the digital age, information and knowledge are valued as commodities, and the expertise that information professionals offer enables the utilization of information for a variety of purposes: research, education, culture, recreation, decision-making, and bibliographic or organization functions of information.

Technological advances have changed all of the functions of information. Information professionals must be cognizant of new technologies and their impact on information agencies and their constituencies, so that strategic planning is possible.

Customizing information service for specific user groups is the vital role of information professionals in the information infrastructure. These professionals provide customized service by employing the service cycle described in Chapter 7: diagnosing a need, prescribing appropriate information sources and services, implementing the service, and evaluating the service. First, we'll take a closer look at how the value and use of information has changed as the paradigm has shifted in the digital age.

The Value and Place of Information and Knowledge in Society

Various sectors of the information professions add to the value and place of information and knowledge in society, recognizing similar and differing values, culture, and research methods for their area of investigation; this includes the creation, diffusion, and utilization of electronic resources; and the development of tools and the integration of systems to link, store, and preserve information.

The role of knowledge managers in corporate America may include integrating archival and records management practices into video instruction files for their companies. Those librarians in the academy may provide publishing alternatives and data storage solutions for their researchers, or those involved in the creation of educational resources may make them openly available on the World Wide Web or in open-textbook initiatives. Then there are those professionals building local cultural heritage collections, where historical library materials are taking on a new digital format, improving discoverability and use of primary resources. We can better understand the role of information professionals by examining the "big picture" of how information is created, disseminated, and diffused in society.

Information and Knowledge Creation

How is information created and diffused in modern society? The complexity of information transfer is represented in Figure 9.1, the Creation, Diffusion, Utilization of Information Model (Greer, Grover, and Fowler 2013, 27).

This model demonstrates where creation, diffusion, and utilization fit into the disciplines from the traditional academic environment in higher education. Information is a valued commodity in every aspect of living in our modern society. As technology emerges, early adopters take interest and explore possible uses for those technologies. This then leads to new ways of incorporating technological developments and digital information into aspects of our everyday life. Once adopted, a new technology will spur new wants and desires, driving the demands for the technological innovation and adding to our reliance on digital information uses and emerging technologies.

Figure 9.1   Creation, Diffusion, Utlization of Information

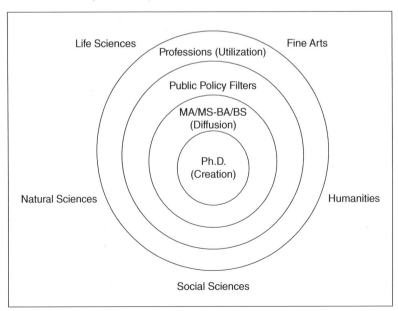

In the information and knowledge infrastructure portrayed in this model, the PhD degree (center of the model) is designed to foster research in order to create acceptable new knowledge for society, knowledge that is within the acceptable parameter of a discipline's paradigm. Although think tanks and private-enterprise research organizations produce new knowledge, that knowledge is often a synthesis or new application of existing knowledge, much of which is produced in PhD programs or by university faculty.

The model shows the traditional divisions of academic disciplines in universities: fine arts, humanities, social sciences, natural sciences, and life sciences. The diffusion circle represents the teaching/learning process that is the central responsibility of the university, the diffusion of knowledge. The outer rings represent public policy and the professions, showing that knowledge is utilized through public policy and its application by the professions.

Public policy is an amalgamation of the laws, regulations, and rules that society requires for stability and structure. Public policy is the result of actions by a corporate body or public agency to enable the use of information. Public policy can support or inhibit information transfer. For example, a governmental body can pass regulations that restrict use of stem cells in medical research or stipulate guidelines for research grants that encourage investigation in certain science areas.

The professions/utilization circle in the model shows that professions promote a screening process that promotes the utilization of knowledge by society. Professions include those schools that teach new theories and methods and support researchers who create new theories and models for practicing the profession, and disseminate new knowledge through journals and professional organizations.

In the digital age, educational programs in private enterprise and online universities have changed the structure of higher education and public education. Although technologies have changed, this model reflects the general pattern of knowledge creation and diffusion in society.

The Changing Information Infrastructure

In Chapter 2 of this book, we discuss the complexities of our changing world and the reality that technology has dramatically altered how we communicate, learn, use, store, and create new information and knowledge. In our fast-paced world, we have come to rely on only one constant, and that is change. We function differently today than we ever have before. In Chapter 3 we examine the historical roots of technologies that have led to societal changes. From Gutenberg's invention of the printing press to a technologically driven environment where information is networked, there is a growing potential for every person to communicate.

Technological advances have fueled ongoing change. We have seen the fundamental perceptions of our Western world reconceptualized as the dominant values are questioned and challenged by emergent paradigms. A new reality has emerged: a holographic, virtual world, where networked systems provide interconnected information. Society increasingly relies on short-term resolution of problems in order to manage the complexities of issues and the uncertainties of functioning in the emergent paradigm.

In Chapter 5 we explored how the information infrastructure has changed the way we live in the digital age. Now, communication is instantaneous; transmitting digitally borne information in our networked world is further challenging us to acquire values that support a different way of thinking and doing. Technology may be understood to be the driving force; however, it is not alone. Other systems are also at work influencing the information infrastructure. The changing factors, or "the environmental and social context," exert a variety of influences on us as individuals as well as on information agencies in society.

At a time when the place of information and knowledge in society supports an ever-changing environment, making sense of information and learning is vitally important. It is in the making of meaning that information is valued and important to our communities. Needed are information experts to aid learners in identifying and utilizing information resources for a variety of purposes. Our information-based economy thrives on learning and then diffusing knowledge.

Libraries increasingly are reaching out to wider audiences by providing services that go beyond the expectations of a print-only environment. The stereotypic image of a library is as the place to go when a person needs a book to read, but that is not the only reason to go to a local library. Public librarians recognize that the culture of communication in our global society is more than the written word. We are increasingly becoming a highly oral and visual society. Information professionals are shifting their values and approaches, modifying the missions of their libraries to break down stereotypes and rebrand themselves.

In order to make sense of information, a community needs leadership from those members who can understand our changing environment. The information professional is uniquely positioned to experiment and engage in the learning process, to understand and then teach and communicate widely, or to make meaningful the vast amounts of information used in our complex society.

The Functions of Information and Role of Professionals

Librarians are information professionals trained to organize, disseminate, diffuse, utilize, and preserve information in ways that provide access for greater public consumption. Information professionals have a vital role in assessing the individual and social factors that influence the utilization process of information. As noted in chapters 6 and 7, a vital part of the information transfer process is diffusion, which enables utilization.

In order to understand what we mean, one must consider the elements of utilization and how library and information professionals are actively engaged in utilization. As noted in Chapter 7, there are six functions of information and knowledge that an information professional works with:

- Creating new information or knowledge: the research function
- Teaching and facilitating learning: the educational function of information and knowledge
- Appreciating: the cultural function
- Enjoying: the recreational function
- Deciding: the informational function
- Organizing: the bibliographic function

The reader may wish to review Chapter 7, where we explained in more depth each function and the reality of the information professionals' passive, reactive, and assertive levels of service. Here we expand on the library and information professional's role with these functions by exploring current examples of professional practice.

Research

The research function creates new knowledge through a formal process involving a review of the literature, framing what is known of the area around the research questions or hypotheses, a plan for data collection, analysis, and a report on the findings. Information professionals may engage in building oral history projects, contribute to the research process with new knowledge to an area of study, or engage in an effort to provide in-depth consultation with researchers.

While on-demand research and information support at a reference desk has long been an accepted service, today's librarians have added the telephone, e-mail, and newer communication modes such as instant messaging and texts. Academic libraries offer support services for individuals through research consultations (offering in-depth, specialized expertise in a subject), online research guides, and tutorials in an effort to reach users and answer questions when they arise.

Washington University School of Medicine Oral History Project includes recorded interviews and transcripts dating from 1959. These primary resources are openly available on the Web. The interviews include reflections on the history of the medical school, medical practice in St. Louis, Missouri, and developments in the field of medicine. This collection continues to grow with new interviews. Since the research function in this example also supports teaching and learning, there is considerable overlap with the education function.

Not all oral history projects are historical in nature; some capture current opinion. In collaboration with disciplinary faculty, oral history research projects are among the growing number of local research collections in both public and academic libraries. Students associated with a course will utilize the Institutional Review Board's protocol for ethical research standards involving human subjects, provide consents, conduct interviews, and then deposit their interview data into the institutional repository. These interviews are original pieces of research that together make up a unique collection or data set that is stored and widely disseminated by the library and may be supervised by the course instructor.

For example, a faculty member from the College of Education at the University of Wyoming is leading the way for oral histories to be captured and included in the *Not Just a Teacher* collection. This oral history collection has growing national participation as graduate students in education conduct interviews, adding to this oral history archive.

Librarians are curating local research collections and becoming publishers by adding value to foster use through ensuring long-term access, sharing, and data protection for these locally built collections, just as they provide storage and preservation for books. Research collections may include undergraduate research presentations, digital research posters, and quantitative and qualitative data sets in institutional repositories created and managed by academic libraries for their scholars.

The University of Wyoming Wool Laboratory Collection is one such local research collection. It spans over a century and involves materials including scientific research in a variety of areas, including wool scouring and processing, agricultural breeding and nutrition, and the economics of wool to regional ranchers and other constituencies, such as the American Sheep Industry. This collection has recently left the College of Agriculture and is now in the library's Emmett D. Chisum Special Collections, which is involved in a collaborative filming project with Agricultural Extension to create a Web-supported video resource providing vital information about this research collection.

The University of Wyoming Wool Laboratory Collection will be curated by cataloging the published resources, digitizing part of the unpublished materials, and conducting a series of oral history interviews to develop a Web-based multimedia tool that will provide context for the collection. This tool functions as video documentation; it has a storyboard that enables viewers to choose their own learning adventure while moving through the historical story of wool science in the region. Important items from the collection will be scanned, placed together digitally, and openly disseminated on the Web as a public educational resource.

The notion of making information more widely available to all of society's members has gained momentum to the point that there is a movement toward more open availability of information, the "open movement." There is a convergence of open access to scholarly information, providing research data openly, and the development of open educational resources.

A unique culture has developed resulting from the disruption of the Internet, which allows many to publish. As a result, new models have arisen to challenge the traditional publishing model. This challenge can be best understood as the open movement that values research and educational information is made available to the wider public. The open movement has provided librarians the opportunity to take leadership in collaborating with university faculty to create sources that decrease the textbook cost for students while providing additional digital learning tools.

In 2008 the National Institutes of Health (NIH) Public Access Policy was passed by Congress, creating a legal mandate that any peer-reviewed article supported by NIH funding should be made freely available to the public via PubMed Central no later than 12 months after the official publication date. In the most recent North American Open Access Meeting, the dialogue supported the notion of convergence. The open movement is converging with open-access publishing, open educational resources, and open data sets and continues to be discussed on an international scale. Librarians can accelerate the transition to a more open system of scholarship through their role in reclaiming the library as publisher during this changing landscape of a scholarly communication system and publishing. This new role is, in part, the role of scholarly communication librarians, and it is the result of the overlap of information functions, such as research, education, and information.

Collaborations among health sciences librarians, scholarly communication librarians, data management librarians, the office of research, and the grant-funded principal investigators can work together to ensure that researchers are compliant with the NIH Public Access Policy and that their scholarship is deposited into open-access repositories such as PubMed Central. Since 2008, when Congress passed the NIH Public Access Policy, national funding bodies have been requiring publicly funded research to be made publicly available. This requirement is expanded to include data in addition to the published scholarly article of the findings associated with the grants. One component includes a data management plan to accompany the grant application. This plan documents how the data will be stored and made publicly available in perpetuity.

Academic librarians generally have a research requirement with their workload. These information professionals are increasingly publishing their "action research" and sharing their solutions to their professional work, or they may be conducting theoretical research that they participate in alone or through collaboration with other faculty. This type of professional inquiry may be both unique and identified as vital to the wider university plan and areas of distinction.

Educational Function

Using information effectively in an environment where technology is always changing presents some challenges. Yet teaching how to use information in today's digital age is an important function that all types of library and information service agencies recognize. Librarians collaborate with colleagues both within libraries and those stakeholders beyond the libraries. Collaboration often includes planning, delivering, and assessing information literacy initiatives within the larger context of instruction and learning in an organization.

New services are focused on interdisciplinary projects and strategic priorities at the campus level, such as providing information literacy instruction through general education requirements that involve teaching students how to enter the academic culture, such as statistics and basic skills. While there are some research methods that can be taught across disciplines, other liaison librarians may be embedded in higher-level courses where specific instruction regarding research methods and resources can be applied.

Such instruction and programming isn't left only to academic libraries, but also includes public and school libraries, where learning opportunities include basic digital literacy sessions and inviting researchers in to give eScience talks.

Academic and public libraries are joining museums in building makerspaces and a sense of place that is dedicated to problem solving, creation, and collaboration. Sometimes it is for DIY (do-it-yourself) craft projects or for building technological solutions such as coding commands for Lego robots or creating electronic boards where electric current flows. Often it's an opportunity for engagement in many areas that are categorized as STEM (science, technology, engineering, and mathematics) or STEAM (which adds *A* for art) opportunities, where teens meet and learn computer coding. The libraries' services aren't restricted to the space among the shelves, but they also invite members to take computer classes or program bots. Some libraries are cultivating community gardens and offering cooking classes with the produce that is harvested.

In academic libraries the liaison or subject specialist's role is best understood through their joint theoretical understanding of a discipline. Many of these information professionals have second master's degrees in the discipline for which they provide information literacy services and research consultation for their disciplinary researchers and students. These research services are possible due to their additional subject expertise.

Embedded librarianship is becoming widely accepted and implemented on campuses. Academic librarians who hold dual degrees can leverage their theoretical training for instruction and research. These dual-educated liaisons serve a vital role in the educational mission of today's colleges and universities when working with professors instructing students at higher-level undergraduate and graduate courses about the many print and electronic resources; these librarians can also answer faculty research questions, encourage them to think critically about class and research resources, and conduct literature reviews. These librarians are actively engaged across campus, seeking opportunities to foster interdisciplinary collaborations by providing research and information services.

Open educational resources provide a wide variety of tools that support learning and challenge information professionals to reconsider their role as educators. For example, a number of universities have partnered to create an Open Textbook Library, recognizing that many college students struggle with the high cost of textbooks. This textbook project is aimed at faculty and students in an effort to make higher education more affordable while providing the best possible electronic resources.

Information professionals are finding themselves in vital roles as change agents, some traveling new paths where their leadership demands risks within the opportunities to create wider solutions for the information infrastructure. Joanne Budler, Kansas State Librarian, was instrumental in paving the way in the creation of a new eBook access model when she repeatedly rejected a library service agreement. Ultimately she won the right to transfer titles from OverDrive to a new platform (Budler 2013).

The education function can easily be recognized in the role of public librarians as they strive to create programming that is innovative and desired by the communities they serve. It is the libraries' employees that are the largest asset of the 21st-century library, not simply the collections.

Libraries and museums of all types have made the move to improve public trust by bringing agencies into the library for educational talks that will strengthen community-based learning. The Center of the American West has been instrumental in working with scientists and the general public regarding environmental concerns. The museum's director and chair of the board is providing conversations in the middle ground that inform, educate, and bring community members more in touch with each other while facilitating civil discourse (Limerick 2013).

In Adams County, Colorado, the Rangeview Libraries have rebranded to become Anythink. The image of Anythink is one of innovation, where the information professionals who manage each library branch are called "experience experts" and the customers who visit the library are referred to as "Anythinkers." The citizens have been invited to belong to a wider vision, one that invites them to discover that anything they can think, they can find at their library. These changes have instilled a sense of pride and membership. The attention to learning can be recognized in the Anythink trademark where their mission is to "open doors for curious minds" and the evidence can be seen in the revitalized and newly constructed buildings, values and library programs (Anythink 2014).

Anythink Libraries won the National Medal for Library Service in 2010. "The library district has revolutionized its operations in nearly every way, making it easier to use, empowering its staff, and establishing dynamic activities for users of all ages, and the results have been phenomenal," said U.S. Representative Jared Polis, who recommended them for the award (Polis 2010).

Cultural Function

Previous economies were based on agriculture and products; the Industrial Age brought more refined products and manufacturing. While we still rely on manufacturing, knowledge products have gained in both quantity and value. eBooks and iPads disseminate cultural works and can provide the platform for teaching oneself to play piano. Consequently, young teens today who readily use the computer are accustomed to learning music without needing to rely on having the large instrument available.

Technologies and systems enable large groups of people to use a variety of devices to share their lives and create a culture of vast networks for communication. Companies don't use paper storage as they once did; digital cloud storage is accessible via the Internet. Some universities have made the switch to Google for e-mail and use Google products for e-mail, course instruction, and communication, infusing new practices to support learning.

An important element of the digital-age culture is the virtual communities people can create using social media. For some it is more recreational as they create multimedia recordings of their day-to-day living and share widely with friends and family in online platforms, such as Facebook. The definition of "friends" has grown to include Facebook friends, "hot dudes," or others known only online. People "friend" others because they are acquaintances. It is widely understood that some want to have lots of friends; therefore, people "friend" strangers, giving them the opportunity to further investigate the possibilities of new relationships, or more likely their page is more of a window into what the person values. "Friends" is a term no longer used in the traditional sense. But these platforms provide a mechanism for finding someone with passion for the same obscure hobby, minimizing the miles between them in our global, networked community.

Library and information professionals can create an image for their libraries that announces the library as a place to go to engage in thinking and learning. The programs offered are topics that enrich the cultural fabric of a community. Libraries of the 21st century are focused on being places within the community, providing space for a wide array of activities, such as makerspaces. Libraries can create places that feel comfortable for people to create, build, innovate, and collaborate with others in the community.

Technology has enabled the dissemination of communication and the storage of vast amounts of cultural information. Some libraries are digitizing their special collections and historical content and making these collections widely available on the Digital Public Library of America (DPLA). DPLA can be found on the open Web, providing access to millions of items that are connected and searchable by timeline, map, virtual bookshelf, format type, subject matter, and the partner agencies that added the digital collections to DPLA.

DPLA enables open access to digitized cultural heritage collections that can benefit many, including software developers, researchers, and others that use primary resources in new and transformative ways. These collections are rooted in the humanities, consisting of digital objects from literature, music, drama, film, dance, and art that provide a focal point for cultural information services.

Recreational Function

Technology inspires a whimsical side in the human spirit. When a family goes on vacation and takes photos of their trip to the Great Pyramid of Giza, one of the oldest of the Seven Wonders of the Ancient World, they invariably return home wanting to share their stories with a sense of "wow" that begs to be shared. Today's social media outlets provide the networks for sharing vast amounts of information generating a wow factor on Facebook, where a story is told, enjoyed, and passed on. Stories are spread in one form of social media to another via Facebook, Twitter, Snapchat, Pinterest, and other mobile apps.

Snapchat is a form of social media that lasts less than 24 hours and offers visual communication that is more information-rich than text alone. A picture taken can have layers added to it, as one can type a one-line statement and add highlight-colored content by drawing or writing by hand on the image. The user can send an ephemeral image that may include a background, a facial expression for emotion, symbols, text, or a colored drawing that supports the notion that a picture can be worth a thousand words. Snapchat can be pushed out to a personal contact that has access to the image for short-term use.

These examples are not restricted to recreational use. Technology that begins as a recreational information function becomes a mode of communication that ultimately can be used for informational and educational purposes, supporting professional development.

Recreational dollars are spent for enjoyment, including purchase of devices, software to play games or songs, or to simply upgrade to the next bigger phone to watch movies, or to a smaller device to make travel easier. For example, a number of apps are designed for star enthusiasts to use for enjoying the night sky. Now parents of grade schoolers can use it to pick out the constellations, planets, or the Milky Way galaxy for a homework assignment. Beginning astronomers can see the image of Orion's belt outlining and connecting the stars that make up the constellation, making the mythical stories come alive in the nighttime sky. The app could even be used to wow a date on a summer evening stroll. Fairly recently, free apps have become even more profitable due to the paid ads surrounding apps; as a result, more quality apps are available for free and supported on a variety of devices.

In the emergent paradigm, the information functions are overlapping. The function of recreation may also overlap with the information function. In other words, information may be needed in order to plan a road trip during time off from work. One might conduct an information search on Google. One might check in with the local department of wildlife to discover where the fish are biting and then plan lunch by looking for picnic tables along the river on a map from the local department of transportation's Web site. Perhaps an interest in rafting during that same trip would drive one to seek out the level of rapids. This could be done by checking out the flow of the rapids, cubic feet per minute readings, to determine if the waves are too high to go down a section of the river.

Informational Function

Information for decision-making can take many forms. Information professionals included, one could seek out information in any number of ways. Maybe a project in the yard has you puzzled as you troubleshoot a solution for your solar outdoor lamps that light your walkway. You reach for your cell phone to look up hours for a store in town or a number to call and inquire about the purchase of batteries. Since information is a valued commodity, the information professional has a valuable role of identifying and synthesizing information for other professionals; for example, in business, law, medicine, and other professions.

As systems have become more sophisticated, new technologies provide new types of information. For example, technological advances have benefited the medical community. A physician can visualize the small bowel, to detect abnormalities without sedation or invasive procedures, with Pill-Cam Capsule Endoscopy, which may someday be used to monitor and diagnose gastrointestinal tract disorders (Given Imaging 2014).

Information professionals are using smart-phone applications and social media to showcase the content of their collections, to announce and widely disseminate the digital collections that are openly available in their repositories, to communicate the library's programming, or to read and/or contribute to the Twitter stream of a national conference, providing valuable insight for colleagues who were unable to attend.

Information professionals provide information for decision-making after diagnosing information need, identifying an appropriate information source, and repackaging that information to address the information need of the client. The client may be a manager in factory, a school principal, a university faculty member, an attorney, or a nurse. Technology facilitates the search for the information and the repackaging of the product into a format suitable for the client and intended purpose. While technology can expedite the search process and enable shaping the information into an appropriate package for instant delivery via the Internet, the key to the success of the information function is the work of the information professional.

Bibliographic or Organization Function

The library is traditionally known for its strict use of subject headings that identify books in the catalog. Such indexing terminology is in a state of flux as clientele become more involved in contributing natural language tags. Some have come to understand this level of public involvement, also known as "crowdsourcing" techniques. One example involved an online puzzle video game where gamers contributed to the biochemistry and structuring of a protein. In the emergent paradigm, the static cataloging system with vocabulary-controlled subject headings is replaced by a more fluid system driven by technological advances. This can be seen in institutional repository software that enables text documents to be full-text searched by the Web crawlers picking up terms matching the words used in a search.

The Digital Herbaria database was created by the National Park Service and the University of Wyoming to organize and make publicly available high-resolution digitized plant specimens. This vascular plant herbarium collection was created to handle the unique need of identifying plant specimens and is built to meet Smithsonian best practices, which require a ruler and color chart placed near each specimen during the imaging process and the inclusion of standard metadata known as DarwinCore. Often a digital collection will require several layers of metadata in order to make the data more discoverable.

In the emergent paradigm, original cataloging and cataloging departments are also switching their workflow to support the need for metadata that enables retrieval of digital collections. The role of these information professionals involves a focus on popular metadata standards, workflow issues, the selection of digital asset management tools, and the metadata schema that are necessary for managing research data. These organizers of information in the digital age discuss metadata-related issues in order to make decisions that support the building of local research collections, adding to the increasing opportunities for information professionals to collaborate across agencies or institutional departments.

Emergent Paradigm Functions of Information

The onset of new technologies has many people changing how they communicate. Information professionals who once relied on correspondence through the mail (U.S. postal service, overnight mail, and e-mail) have benefited from the increasing speed of communication, improving the productivity of many information professionals. With the reality that communication can be sent and received so quickly, there has been an increase in decision-making and project results. Some will even send a text to alert a colleague to check their e-mail so that the response rate can be expedited. Others are using social media to spread the word about their individual scholarship or organization's work.

As the complex nature of our society increases, the functions of research, education, culture, information, recreation, and organization of information have continued to blur in the emergent paradigm. Overlaps among the functions are occurring. What does this blurring look like? Imagine watching a YouTube video and learning to become a better chess player, then challenging Uncle Joe to a rematch the next time he is in town. This example is a cross between the education function and the recreation function. The lines are blurred.

For a large-scope example, let's examine the century-old University of Wyoming Wool Laboratory Collection that was housed in the College of Agriculture until the donation of its papers, publication library, resource materials, and fleece samples to the university library system. The Wool Team put together a plan to process or curate the donation by working across library departmental lines. The content in the wool collection overlaps information functions.

If we look at this collection from the perspective of the research function, we recognize that it includes primary source materials that researchers in a variety of disciplines can use. They may use the wool science that includes the processes for cleaning fleece and shrinkage that occurred, the nutrition and breeding science that took place over many decades, and the possible new knowledge from the DNA in the jars of wool fleece that the American sheep industry deems valuable as a historical record for sheep breeds from the samples of sheep flocks throughout the world.

When thinking about the collection through the perspective of the education function, the notion of overlap becomes clearer. Many of these primary sources were used for instructional purposes for the wool science program, providing education on such topics as animal husbandry, fabric and fiber uses, the business of ranching, and the economics of wool. When many of these educational resources become digitized, the wool collection can become openly available for other educators. When available digitized, K–12 school-children learning about history or preparing for a History Day competition and needing primary resources (education function) have information; so do 4H clubs working with sheep or wool fiber arts and crafts (recreation), ranching and the Basque influence in the community (cultural function), or stories about life in the West (cultural) or the cost of wool (informational); so do the hobby farmers who didn't grow up on a ranch but now find themselves raising sheep or selling wool (education and recreation).

As information and knowledge are used to teach or learn, enjoy, appreciate, and understand our world, make decisions, create new information or knowledge, and to find or locate information, there is an increased blurring of these functions. By shifting the perspective to a different function, a different insight is provided, making the overlap more readily identifiable.

These six functions involve many moving parts, coalescing and separating; consequently, the professional's role with information and knowledge becomes more dynamic than ever before.

The work to make the collection organized and accessible on paper is a challenge, but to digitize the collection and increase the opportunity to have it found on the open Web will continue to open the collection to a variety of uses. The information infrastructure in the emergent paradigm is nonlinear, and the values of the emergent paradigm remain undefined in its flexibility, unpredictability, and fullness.

Summary

The emergent paradigm breaks through traditional barriers and definitions, providing an automatic opportunity for information professionals to embrace their roles as leaders in the digital age. Learning is taking place in many aspects of our day-to-day living; although we have identified the functions of information, today's use of the information infrastructure helps those who are using information for many purposes to better understand their work and view it from a broader perspective. An educated information professional has the opportunity to become an innovator, a collaborator, and a leader during this time of change by embracing the information infrastructure and by being aware of the elements of the information transfer cycle.

An example of the emergent paradigm is the new position of the academic library change agent, the scholarly communication librarian. This librarian is hired to bridge the traditional library to that of the 21st century. The objective of the scholarly communication librarian is to support a growing research culture, to collaborate, and to bring in service agencies, administrators, and researchers across campus in support of research and development solutions that support the building of a scholarly repository and a scalable data infrastructure that enables campus researchers to apply for national research grants. Once the infrastructure is in place, then cohesive data management plans for data storage, backups, access, and dissemination will provide additional benefits for project researchers. It is an exciting time for libraries and for the information professionals who provide the leadership.

References

Anythink: A Revolution of Rangeview Libraries. "Jobs." http://www.anythinklibraries.org/job-center/jobs.

Budler, Joanne. 2013. Beta Phi Mu Beta Epsilon Induction Ceremony Keynote Speaker. Emporia State University School of Library and Information Management. Online presentation, May 5.

Given Imaging, Ltd. 2014. PillCam Capsule Endoscopy. http://www.givenimaging.com/en-us/innovative-solutions/capsule-endoscopy/pages/default.aspx, accessed October 25, 2014.

Greer, Roger C., Robert J. Grover, and Susan G. Fowler. 2013. *Introduction to the Library and Information Professions*. 2nd ed. Santa Barbara, CA: Libraries Unlimited.

Limerick, Patty. 2013. "Understanding the Underworld: Hydraulic Fracturing and the Depths of the Humanities." Presentation at Wyoming Institute for Humanities Research, Laramie, October 30.

Open Textbook Library. http://open.umn.edu/opentextbooks.

Polis, Jared. 2010. "Anythink Libraries Receives Nation's Highest Award for Community Service." November 16. http://www.anythinklibraries.org/news-item/anythink-libraries-receives-nation%E2%80%99s-highest-award-community-service.

University of Wyoming College of Education. "Not Just a Teacher: Oral Histories." Wyoming Scholars Repository. University of Wyoming Libaries. http://repository.uwyo.edu/njat_oralhistories.

Washington University School of Medicine Oral History Project. Becker Medical Library. St. Louis, MO: Washington University. http://beckerexhibits.wustl.edu/oral/interviews/index.html.

10
Emerging Trends

Chapter Overview

In this chapter we review major concepts introduced earlier. As the digital age has emerged, the information transfer processes have changed, and these changes are examined as trends in the information infrastructure with implications for library and information professionals and for library and information professional education. The chapter concludes with a glimpse into the future of the information infrastructure.

Summary of Important Concepts

In this book we have provided a framework for understanding our complex, confusing, and emergent digital world. We provide the reader with a lens for examining, in a comprehensive manner, our information infrastructure. We also empower the reader to understand how technology, through the Internet and social media, has changed communication patterns and our 21st-century culture.

Information Transfer Model

The framework for understanding the vast, global information infrastructure is the information transfer model: the creation, recording, mass-production, dissemination, organization, diffusion, utilization, preservation, and discarding of information. The reader may wish to review a detailed explanation of this model in Chapter 4.

Environmental Context

Each of the stages of information transfer is influenced by environmental factors: culture, physical geography, political structure of society, legislation and regulations, information policy, the economic system of society, and technology. These factors are described in Chapter 2.

These environmental factors interact to impact society, causing change, which may be gradual and incremental, or it may be sudden and cataclysmic. Even gradual change can result in major changes in attitudes and values, a paradigm shift or change in how people view their world. That paradigm shift is described below.

Influence of the Emerging Paradigm

In Chapter 2 we provided a model to assist in understanding the networked digital world that now engulfs us; the work of Peter Schwartz and James Ogilvy (1979) and their analysis of theory building in different disciplines produced a model that explains the shift to the emerging characteristics of the new paradigm. This model describes the collapse of certainty in society and compares the dominant and emergent paradigms, which we summarize below:

> **Dominant > Emergent**
> Simple > Complex
> Hierarchy > Heterarchy
> Mechanical > Holographic
> Determinant > Indeterminate
> Linearly causal > Mutually causal
> Assembly > Morphogenesis
> Objective > Perspective

This model provides a framework, a way of thinking about the changes occurring in our society and the implications for changes in the information infrastructure. The role of the information professional has become more complex because of the transition to the emerging paradigm—because the traditional hierarchical structure of organizations is often morphed into a flatter organization, a heterarchy with multiple orders, and the traditional lines of authority are replaced by more autonomy within the organization. Rules may be changed or eliminated, and the change to a new organizational structure can cause strife within an organization.

New technologies and new systems for communication change the way people communicate. Information professionals, as leaders, must be aware of these changing values, structures, and communication patterns, because information professionals are key facilitators in the information infrastructure.

Competing Paradigms in the Information Professions

As paradigms have shifted in society, so has a paradigm shift occurred in the library and information profession. That shift is from a "bibliographic" to a "client-centered" paradigm.

The bibliographic paradigm is so named because the preoccupation has been with books and other materials and their organization. The library profession valued the acquisition, collection, organization, and storage of the records of society and access to them. Collection size was valued: the largest collection of materials, the largest library building, the largest staff, and the largest budget were considered the best.

Organization of those materials became a priority. Although collections expanded after World War II to include media in a variety of formats, the concern for organization and care of the information packages was at least as important as the concern for the users of information.

As we entered the digital age and technology became more sophisticated, the preoccupation sometimes shifted to applying technology, having the latest computer systems, and a focus on "access." Although innovation is important and the newest technologies should indeed be incorporated into information services, the needs of the user are sometimes secondary to maintenance and utilization of the technology instead of focus on the information users.

Diffusion of information and knowledge requires helping a client to establish meaning. As we have stated in this book, diffusion of information and knowledge is the goal of library and information professionals as guides and facilitators in the information infrastructure. The "people paradigm" replaces the focus on a collection of information and technology with a focus on people. This is a relatively simple concept but one that is often overlooked in library and information science (LIS) schools and in professional practice. Recognition of this shift is important for library and information professionals to be effective guides in the information infrastructure.

Library and information professionals must also be aware of the impact of the emerging paradigm on information services. Many changes are under way, and major trends are outlined below.

Role of the Information Professional in the Information Infrastructure

As explained in Chapter 9, information professionals play a vital role in the information infrastructure because they grasp the "big picture" of the information infrastructure. They understand the information transfer processes, from creation of information to recording, mass-production, dissemination, organization, diffusion, utilization, and preservation or discarding.

Information professionals understand how technology can change each of the information transfer processes. The traditional single author gives way to multiple voices, fast change, and lack of control. The Internet and social media eliminate the ability of individuals and governments to control the exchange of ideas. In the past, new technologies like Gutenberg's printing press facilitated the mass-production of information and new knowledge to larger audiences. Now the creation and dissemination of information can circumvent the former gatekeepers—publishers, broadcasters, and media producers—so that individuals with smart phones, tablet computers, or laptops can create information packages and disseminate them widely through social media immediately. The same media enable instant feedback by the audience. Barriers of time, distance, and cost have been eliminated or drastically reduced.

The models mentioned above are synthesized in Figure 10.1. The environmental context influences the prevailing paradigm in society, and the interaction of paradigm values and variables in the environmental context all influence each of the stages of the information transfer processes. The arrows and ellipses indicate the interactions of these variables in a very complex interaction with information transfer, the information infrastructure. These trends are part of a much greater wave of change, as described below.

Figure 10.1   Information Transfer

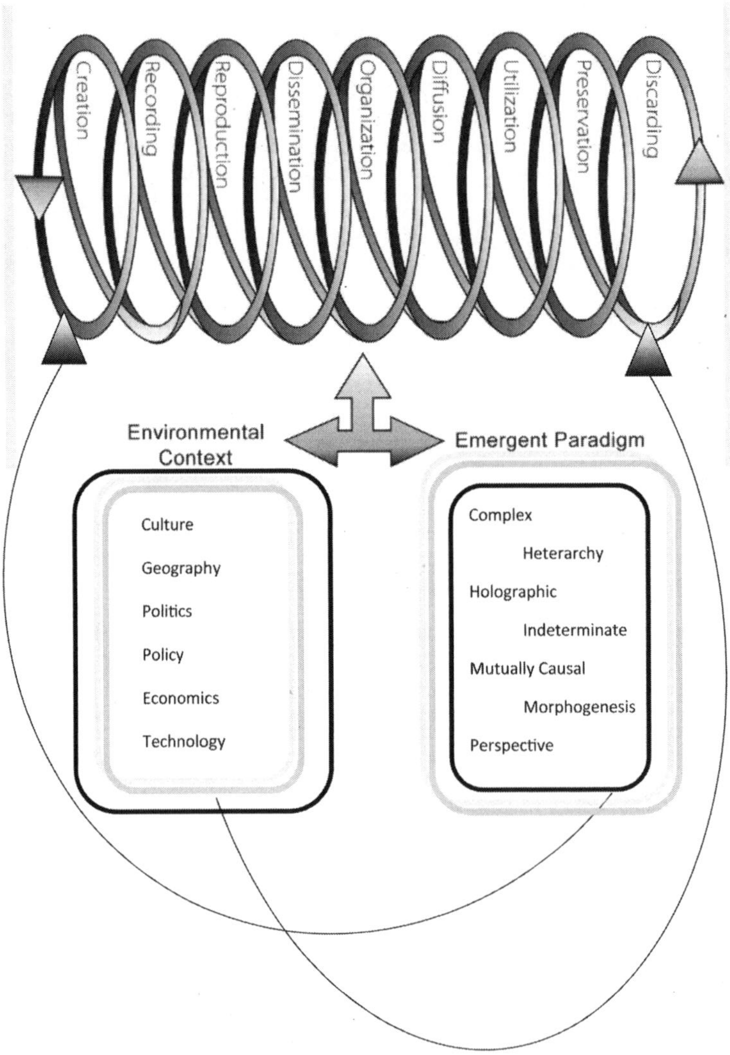

The spiral labels read: Creation, Recording, Reproduction, Dissemination, Organization, Diffusion, Utilization, Preservation, Discarding

Environmental Context

Culture

Geography

Politics

Policy

Economics

Technology

Emergent Paradigm

Complex

Heterarchy

Holographic

Indeterminate

Mutually Causal

Morphogenesis

Perspective

Emerging Trends in Information Agencies

The library and information profession traditionally has structured professional organizations by type of library; for example, the Public Library Association within the American Library Association (ALA), College and Research Libraries Section of ALA, Special Libraries Association, and American Association of School Librarians. However, just as technology collapses barriers of distance and time, so does the emerging paradigm collapse traditional organizational structures and practices. Consequently, we will discuss trends generally, noting some applications within types of libraries and information agencies.

Globalization

The Internet transcends geographical boundaries, and English is evolving as the international link language. Global partnerships are emerging; for example, the electronic library in Ukraine is funded by international organizations, including the Gates Foundation, and publishers, especially European publishers.

Through the Internet, local library catalogs have global reach. When a public library or university catalog is linked to WorldCat, a library user can identify other libraries that have a certain work that may not be available in the local library's collection. Through interlibrary loan, a client can retrieve items within a few days, and the local collection is thereby expanded beyond state and national boundaries.

The information infrastructure is as close as one's smart phone or laptop computer, which is global in reach. People in developing countries have access, as do people in developed countries.

Open Access

The open access movement, which is international in scope, is revolutionizing the information infrastructure, and library and information professionals are leaders. The Budapest Open Access Initiative in December 2001 was a conference convened in Budapest "to accelerate progress in the international effort to make research articles in all academic fields freely available on the Internet."

The Berlin Declaration on Open Access to Knowledge in the Sciences and Humanities is another major international statement on open access that was developed at a 2003 conference hosted in Berlin. The objectives of the open access movement are to (1) maintain peer review standards of quality for research while (2) making journal articles and other academic publications available free for access and reading, yet published with some other means of recovering cost (e.g., subsidies or charges for hard copy publications), and (3) modifying traditional copyright practices so that academic work can be used more freely as a foundation for additional research (Berlin Declaration 2003).

The open access movement challenges the traditional model for academic publishing. This traditional model may not be sustainable with the widespread availability of technologies that enable the recording and dissemination of information in new ways.

Repositories Change the Librarian's and Client's Roles

Establishing a repository is an element of implementing open access. A repository changes the behavior of both faculty or other clients and librarians. Librarians and clientele must work together in different ways. They must collaborate much more so that the client is aware of the issues involved with storing research data and information and organizing it, and aware of possibilities for disseminating research results. The librarian can be a valuable partner in this process. In the traditional research and publication process, an individual researcher or a research team could work independently as they collected, stored, and analyzed data. They wrote their research report and sought a journal or academic press to publish the results. The librarian usually wasn't consulted.

With a repository on campus, the librarian can be a valuable part of a research team, to provide guidance and assistance in the organization and storage of data in a repository. The librarian can assist in the processes of preparing a document for dissemination and by clarifying copyright issues and protection of the researcher's creative work. The librarian can help locate suitable commercial or academic presses for wide dissemination of the work.

Establishment of a repository also can widen the dichotomy between research and teaching institutions. In a research institution, research and publishing is an inherent and expected activity. In an institution that emphasizes teaching, research is a lesser priority, and teaching loads are heavier. Teaching institutions may assign faculty to twelve hours of classes each semester, and a research institution may assign only six hours, and faculty are encouraged to obtain grants to hire adjunct faculty to teach part or all of their classes. In teaching institutions, faculty do the minimum amount of scholarship in order to attain promotion and tenure, and their role may be to utilize the repository as a teaching tool rather than as a research tool. Students may be taught to access the local repository to read the products of university students and faculty, and they may be encouraged to deposit their major papers and theses in the repository.

Regardless of the type of institution, the repository can be used to further the university's mission, and it requires change in the way that faculty, librarians, and students use information and fulfill their professional roles. Repositories can enhance scholarship and opportunities for discovery.

Repositories are not restricted to colleges and universities. A repository is a storehouse for knowledge that's created locally. Public libraries are now looking at how they can support people developing their own content, whether print, video, or multimedia.

The "maker movement" is all about the library supporting people as they create information packages. Although libraries in the past only collected and disseminated information packages, now libraries of all types are supporting the creation of information packages.

Outreach

A trend in all types of libraries and other information agencies is outreach. Previously outreach in a public library was to do a book talk in a school, or an academic librarian would guest lecture in a class. Now in academia the specialist librarian has office hours in an academic department and develops ways for students to engage in research as part of their course work. Academic librarians engage with faculty in their research as well, as described in Chapter 9.

In public libraries outreach has expanded considerably. For example, librarians in rural areas are doing such things as setting up a library booth at the farmer's market in their communities.

The Douglas County (CO) Public Library pioneered the community reference model that placed librarians in community organizations and government offices. Librarians become part of community organizations dealing with important issues, and the librarian does research and writes reports to help guide decision-making on local issues in the community.

Public library programs are expanding beyond the traditional story time and offering various kinds of literacy-enriched and learning-enhanced programs: cooking classes, model gardens, gardening areas for people to plant their community gardens, nutrition, yoga classes, and anything that supports community well-being. These programs expand our notion of libraries supporting learning, especially in public libraries, where the definition of "learning" includes learning to improve one's quality of life.

From Transactions to Engagement

The activities of library and information professionals are changing as the emergent paradigm influences the profession. As noted above, repositories change the roles of information professionals, and outreach efforts have increased. Also, the interactions with clientele now require more engagement. To address a client's information needs, a diagnostic interview is necessary, along with questions to identify the person's actual need and to identify appropriate information sources that will fulfill those information needs. As a result, information professionals are engaging more with clients in the library and leaving the library to engage with the community.

Information Literacy and Metalearning

"Bibliographic instruction" has evolved into information literacy—teaching students to identify, organize, evaluate, and present information. Many libraries are moving toward teaching "metaliteracy" so that students learn about the plethora of avenues for production of information, learning to effectively create information using traditional media as well as social media.

Information users of all types, especially students, deserve to learn how to do research efficiently, and browsing is not the only strategy for finding information. It is the role of library and information professionals to take a leadership role in teaching clientele to locate and evaluate information. Information consumers must be wary of content when consulting information sources in any format.

Electronic Publishing

A major trend is the move to electronic publishing and what that means for library collections. This trend suggests that libraries will rent more and more content instead of owning it. By giving up ownership, libraries will have little control over what happens to that content. Traditionally, the library's role has been perceived as keeper of content for the public, to ensure that the public would have access indefinitely. In the digital age, what is the librarian's role in ensuring what content is preserved and not censored?

Some university presses are partnering with the library, adding publishing as a collaborative activity. Thus they are connecting the publishing process to the university repository.

From Objective to Perspective

A hallmark of the 21st century is the collapse of certainty in society. The intellectual revolutions that took place in the fields of physics, chemistry, and biology began the destruction of the Newtonian image of the machine world. In a world of multiple perspectives, the role of the information professional becomes ever more challenging. The notions of objectivity and truth depend on one's perspective. People tend to believe what they want to believe, and reading can support or challenge understandings; readers must be critical of the sources they cite.

The blogosphere has been described as an "echo chamber," and people tend to seek out those blogs, Web sites, radio stations, and television stations that support their views. In recent years, many mass-media news sources, including television, radio, and newspapers, have abandoned a dedication to objective news reporting to report news with a point of view. On television Fox News leans to reporting with a Republican or conservative perspective, while MSNBC espouses a Democratic or liberal perspective.

Information professionals must be aware of the perspectives espoused by news media, publishers, organizations, and leaders. When helping people during an information search, information professionals must be alert to multiple perspectives on issues and inform clients when ideologies are encountered. In this digital age, multiple perspectives must be recognized and valued.

As we explained when we discussed diffusion (Chapter 6), the role of an information professional is to support diffusion and utilization, which means that we must do what we can to help people understand the information and knowledge that they encounter. This is a vital task of the information infrastructure.

Technical Services Are Outsourced

The cataloging of books and maintenance of a card catalog was a major activity of libraries before technology invaded our world. Large libraries maintained a staff of professionals and paraprofessionals to catalog books, assign subject heading, prepare catalog cards, and file the cards in the catalog. Gradually, this cataloging operation was replaced by "copy cataloging"; a library staff member (often a paraprofessional) is trained to locate cataloging information and to download it. Very little original cataloging is done, even in the largest libraries. We find the information that we can download from a book jobber. Few libraries now maintain a position called "cataloger." However, a catalog is essential for defining the location of a library's resources. The metadata librarian is the cataloger in the digital age.

Technical services could make a difference if users can add their own tags to books. Perhaps library catalogs should be viewed as a social medium. A client could add a tag on a local level, even though it may not be a Library of Congress subject heading. Information professionals should be willing to enable clients to have input in catalogs and other guides to collections as a way of encouraging diffusion.

Lifelong Learning

The diffusion of knowledge is a vital function of library and information professionals. Therefore, the library is a lifelong learning unit in any institution or community. A student completes a major to show that she can do work in depth. However, a student at any level should be charged with learning *how* to learn, in order to be equipped for lifelong learning.

Children's Services

To be relevant in the future, libraries must provide services to children. Over the last ten to fifteen years, there has been a focus on early childhood services, and that trend continues to expand. Sharon Morris, the director of library development at Colorado State Library, said the following:

Research on literacy shows that the first three years of life are critical. Children's librarians are being vigilant and proactive, learning all they can about brain research and child development and changing story times by going beyond a fun experience for kids—inspiring them to read but also providing an educational experience for children, parents and child care people.

Librarians now provide parents with suggestions for home activities they can share with their young children. For example, some librarians give cards to parents for things they can to do at home with their children. Parents are encouraged to be the first teachers in children's lives and to assist them as they take their first steps along the path of lifelong learning.

Trends in Schools

An ongoing trend in schools is the establishment of standards for learning and the assessment of students' learning. Teacher evaluations are also aligned with the evaluation of students' learning. At the national level, Common Core has been established as standards for science, math, and English/language arts. These standards have been adopted by 45 states.

The central and logical person for standards implementation is the school librarian, because he or she works with all school personnel and knows curriculum. The current standards for school librarians, *Empowering Learners* (2009, 8), states that the mission of the school library media program is "to ensure that students and staff are effective users of ideas and information. The school librarian empowers students to be critical thinkers, enthusiastic readers, skillful researchers, and ethical users of information." The school librarian is in a position to be a leader in the school and central to the mission of the school in the 21st century.

From Bibliographic Paradigm to Digital

The library in the bibliographic paradigm was a warehouse of books and materials; Morris (2014) notes that "now libraries are being designed for people to have experiences and activities and engagements."

LIS education in the past taught bibliographic functions: how to collect, organize, and disseminate information. Schools of library and information science also taught the importance of preservation. Now technology enables information users to tag documents the way they want to. They can tag articles, books, or Web sites for their future use. They come to understand that they don't yet have the language to talk about it, but authority control is important. If different words are used for the same subject in indexing, access and retrieval are hindered. It's now possible for users to add tags to the library's catalog.

Organizing information and helping people to locate it have been and continue to be important functions of information professionals. As technologies change, the way that the bibliographic function is performed also changes.

From Authority to "Crowdsourcing"

Since the beginning of the profession, library and information professionals have assumed the role of authorities for organizing and evaluating information resources. Members of the profession have been instrumental in creating review sources and reviewing various media. In this digital age, the responsibility for evaluating information is with readers, sometimes called "crowdsourcing." Anyone can create videos or books, add to Wikipedia, make a Web site, and otherwise create an information source. Readers or viewers of these information sources also have the opportunity to critique the information. The library and information profession can no longer ensure "authority" in the same way as in years past. Instead, authority is shared with information consumers.

The Challenge of Remaining Relevant

A trend for all information agencies is the need to remain relevant to the people they serve and to the people who provide them with funding. Information professionals must demonstrate added value that the librarian provides for the community they serve. For example, the corporate librarian must make a case for the databases, including how people are using them, tracking resources, and providing corporate intelligence.

To survive and thrive, librarians must be proactive and creative as they gauge the needs of the organization and the community they serve and strategically reach out to address those needs. Demonstrating the impact of libraries is being done more effectively than ever before, by providing the research and reports to funding bodies and administrators to let them know the impact the library is having.

Leadership in the Profession

Library and information professionals must provide leadership for the advocacy of libraries and information services, especially in times when state and national leaders advocate for "small government." That term usually means less money for public services, including education, libraries, and information resources, which are critical components of the information infrastructure. Library and information professionals must be assertive leaders, actively engaged in political activities to advocate for these personnel, resources, and services that are essential for a knowledge society in the 21st century.

Currently there is no government agency responsible for leadership of the nation's information infrastructure. The National Commission on Libraries and Information Science during the years 1970–2008 was such an agency and was consolidated into the Institute of Museum and Library Services (IMLS), an independent agency in the U.S. government. The mission, displayed on the IMLS Web site, "is to inspire libraries and museums to advance innovation, lifelong learning, and cultural and civic engagement." This mission is much too restrictive to provide leadership for the national information infrastructure.

Preservation Concerns

Information professionals are responsible for keeping and providing access to the cultural records of a society, and a growing percentage of records is now available in digital form. Traditional letters have been replaced by e-mail and text messages. Books are available in both paper and electronic form.

As technologies change, how will our society store the research, literature, and other art forms that traditionally have been stored in a paper format? Currently the decision to digitize books is driven by the market-place; publishers will convert to digital format those titles that are expected to be profitable. Yet in order to preserve works for future generations, we as a society will need to address this issue centrally, through some govern-mental agency, and the information profession should play a leadership role. Preservation is a key issue for the future.

Implications for LIS Education

The intellectual revolution that has taken place must be addressed in library and information science (LIS) education. The metaphors of the emer-gent paradigm are altering how we understand the global society, govern-ments, science, individuals, and human conditions. Individuals, groups, and society generally are affected by this shift toward a more complex reality. The challenge for LIS education is to see the world, professional organiza-tions, information agencies, and technology with fresh eyes. Innovation and change are constants bringing about a world of unintended consequences and unpredictable situations.

LIS schools must teach students to be alert to societal changes and to lead change in library and information services. More than ever before, information professionals must be schooled in the important theories and concepts undergirding the profession and the functions that the profession performs. Students must be educated to see the "big picture" of the role of information in society and how it is played out in the global information infrastructure. Students should understand human information processing, information transfer, information engineering, and information organiza-tion management. Information technology must be integrated into learning about each of the information transfer processes, just as technology is inte-grated into all aspects of our lives.

Leaders must be able to identify trends and to lead their staff to change services and facilities to address these trends; therefore, LIS curricula must incorporate "big-picture thinking" as well as applications of demonstrated principles of information organization, dissemination, and utilization. Cur-ricula must include theories, principles, and applications, including the opportunities for students to practice their newly acquired knowledge of the profession. Schools should include the opportunity for each student to have a supervised practicum experience in an information agency, an internship, or the equivalent of student teaching.

Also, professionals in the field and LIS schools must recruit bright, cre-ative, innovative, courageous, energetic students who can accomplish these goals. These prospective information professionals must also possess those personal characteristics that accompany effective leadership—especially a positive disposition and self-confidence.

LIS schools traditionally have been able to recruit people who uphold the values of the profession, but the value of openness to change has not been as prominent. In the curriculum, there should be more of a focus on how research can drive methods. More research-methods classes are needed in order to teach graduates how to assess services offered and processes.

As one library leader stated, "Knowledge competencies are becoming less important than disposition competencies" (Morris 2014). Children's librarians must be able to track the latest research on child development and brain development. They must also be able to talk to parents. Library education must prepare people to make decisions based on community needs identified by evidence. Librarians must also be able to diagnose the information needs of individuals and read nonverbal communication cues.

Evaluation is necessary to determine which services to maintain, which services to discontinue, and which services to add. Information professionals must be intentional about testing their effectiveness, and LIS education must include instruction in research methods to enable students to make evidence-based decisions.

The Information Infrastructure in the Digital Age

Information professionals are expert on the information infrastructure and utilize a variety of social and information theories to provide leadership in the creation, reproduction, dissemination, organization, diffusion, utilization, preservation, and discarding of information and knowledge; therefore, library and information professionals can diagnose information needs, then provide services and sources that address those needs. These services include the gamut of information functions: education, information, research, culture, recreation, and organization.

Throughout this nation, public agencies face cuts as local, state, and federal governments are challenged to reduce services. Information professionals must make themselves and their information agencies essential in order to survive and prosper. To do so means changing with societal trends in order to meet the needs of our communities.

Libraries and information agencies must partner with their communities, whether local, national, or global; library and information professionals must be engaged in their communities as leaders, building trust and establishing relationships, to survive and thrive. Community involvement requires being proactive, willing to take risks, and armed with an entrepreneurial spirit.

Through the leadership of information professionals, the information infrastructure nourishes a society through the information transfer cycle. The stages of the information transfer model remain the same, but the ways that each process is manifested in the infrastructure have changed significantly in this emergent paradigm prompted by digitization. Table 10.1 portrays these changes.

The digital era is best described as "complex." Each of the information transfer processes is more complex and requires the information user to make choices; consequently, the role of information professionals is also more complex, because they must be able to diagnose users' information needs, help them locate information resources in appropriate formats according to their preferences, and assist them in interpreting and making meaning of the information.

The relative simplicity of the bibliographic paradigm has been replaced by an emergent paradigm that requires a much more challenging role for information professionals. More than ever before, the information professional and the information consumer are required to use their imaginations in order to partner in the acquisition and effective use of information.

In other words, the information infrastructure has evolved in ways described by Schwartz and Ogilvy. These ways move from simple and organized to complex and fluid, from analog to digital, from hierarchical structures to heterarchical, from determinate to indeterminate solutions, from causes that are linear to mutually causal, and from objective viewpoints to perspective views.

Table 10.1   Changes in the Information Infrastructure

Information Transfer Process	Bibliographic Paradigm	Emergent Paradigm
Creation	Individual researchers use quantitative, objective methods; conduct practical research	Teams of researchers use a mix of methods, apply theory to professional research
Recording	Research is recorded and stored privately; data sets may be discarded after publication of results	Research is recorded digitally and stored in a repository open for review; data sets are preserved
Mass-production	Reports are peer-reviewed, submitted to printed scholarly journals, copyrighted	Reports are peer-reviewed and accessed through a repository, may be submitted to scholarly journals; work may be Creative Commons licensed and possibly copyright registered
Dissemination	Journals published as hard copies and sent to subscribers; work is reviewed in publications	Repository is open-access; journals are disseminated to subscribers digitally, assessed in social media
Organization	Journals are organized in libraries using standard cataloging and subject headings; metadata are limited to a brief description of the content and location of the physical document	Repositories create customized tags, and users also tag with links to other sources; metadata elements are greatly expanded to locate other information sources
Diffusion	Education occurs primarily face to face, using technology to augment instruction	Education occurs in a variety of technologies, often at a distance
Utilization	Utilization occurs with individual assistance and self-learning	Utilization occurs with assistance via technologies and social media; usage can be determined
Preservation	Paper is the primary medium	Paper is less used; digital sources are the primary storage medium
Discarding	Discards are physically destroyed or given away	Digital information resources are deleted

The emergent paradigm requires an imaginative approach in order to address the complex issues that arise. Old solutions don't work in this emergent-paradigm digital world. Imaginative approaches will enable us to apply the information infrastructure in new and creative ways to solve problems.

The Information Infrastructure in the Future

"Future studies" is a recognized field of study, and the World Future Society publishes the *Futurist*, a journal that enlists futurists to forecast global trends. Another source of societal trends is the annual New Media Consortium Horizon Report, which compiles the forecasts of international experts on technology trends and the impact on society globally; one such report centers on trends that impact libraries (Johnson et al. 2014).

The American Library Association, with funding from the Institute for Museum and Library Services, has established the Center for the Future of Libraries. The purpose of the Center is to identify emerging trends relevant to librarians and libraries and to collaborate with innovative thinkers to help library and information professionals address emerging issues.

Since the present is indeterminate in the digital age, so, of course, is the future. We'll leave forecasting to futurists and technology experts; however, we can propose questions that lead us into the future. Following are some questions that we invite the reader to consider:

- Is there a need for an international policy-making body to oversee the information infrastructure? We live in a global society in which countries and their governments are interconnected; however, will an international agency only obfuscate infrastructure issues instead of applying imaginative and futuristic emergent paradigm approaches?

- Who should advocate for the information infrastructure? How?

- Should leadership for the information infrastructure in the United States come from a person within the government structure who assembles a diverse, knowledgeable committee or task force to plan strategically? Or should leadership emerge organically from multiple stakeholders?

- Given the open access movement, should copyright be restructured? If so, how? What are the global initiatives and policies that should be considered?

- What are the implications for library and information science education in the digital age?

- Research on dominant issues related to the information infrastructure are discussed and the subject of articles in professional journals. How are such issues reflected in LIS curricula?

- Do the library and information professions and LIS education have the necessary leadership to lead the information professions into the indeterminate future? Where will visionary leadership come from?

- Who and what organizations should provide leadership and supervision of the information infrastructure?

- How can library and information professionals remain current?

- What changes should be made in information policy (including copyright)?

- How will the roles of libraries and librarians change?

- How will the repository movement and other innovations be funded to forward the infrastructure? Grants? Government support?

These and many other questions help us to face the uncertain future that stretches before us. The challenges are great and likewise the satisfaction that the future will bring to a profession that is central in the information age.

Summary

Technology has been a driving force in Western culture, and the result has been a fast-paced society that is characterized by complexity and uncertainty. The digital age has prompted the emergence of a new paradigm that has dramatically changed the information transfer processes and the role of library and information professionals. These changes are examined along with trends that are influencing and will continue to influence the information infrastructure and the role of library and information professionals. Although the future is indeterminate, we can be guided by questions that have emerged. The emergent paradigm brings changes and uncertainty, but it also promises a vital role for library and information professionals.

References

American Association of School Librarians. *Empowering Learners: Guidelines for School Library Programs*. 2009. Chicago: American Association of School Librarians.

Berlin Declaration on Open Access to Knowledge in the Sciences and Humanities. 2003. http://openaccess.mpg.de/286432/Berlin-Declaration.

Budapest Open Access Initiative. 2001. http://www.budapestopenaccessinitiative.org/background.

Center for the Future of Libraries. American Library Association. http://www.ala.org/transforminglibraries/future.

The Futurist. Bimonthly. Bethesda, MD: World Future Society.

Institute of Museum and Library Services. http://www.imls.gov.

Johnson, L., et al. 2014. "NMC Horizon Report: 2014 Library Edition." Austin, Texas: The New Media Consortium.

Lakin, Jacqueline McMahon. 2014. Interview by Robert Grover, July 1. Kansas State Department of Education, Topeka.

Morris, Sharon. 2014. Telephone interview by Robert Grover, August 19.

Sheridan, John. 2014. Interview by Robert Grover, June 19. Emporia State University, Emporia, Kansas.

Index

Note: an *f* indicates a figure; a *t*, a table.

About the Authors

ROBERT J. GROVER is retired associate vice president for academic affairs and dean of graduate studies at Emporia State University, where he had held the position of dean and professor of the School of Library and Information Management. He has also held faculty and administrative positions at the University of South Florida and University of Southern California. He earned his MLS and PhD from Indiana University and has coauthored *Assessing Information Needs: Managing Transformative Information Services, Helping Those Experiencing Loss: A Guide to Grieving Resources,* and *Introduction to the Library and Information Professions* (2nd edition), all published by Libraries Unlimited.

ROGER C. GREER, PhD, MLS, was a creative thinker in the public and academic library professions for more than 50 years. He was dean emeritus of the School of Library and Information Management, University of Southern California; former dean at Syracuse University; and professor emeritus at Emporia State University. He earned his master's degree in library science and doctorate at Rutgers University. He is coauthor of Libraries Unlimited's *Assessing Information Needs: Managing Transformative Information Services* and *Introduction to the Library and Information Professions* (2nd edition). Greer's work has also been published in numerous professional journals.

HERBERT K. ACHLEITNER, professor emeritus of the School of Library and Information Management, Emporia State University, earned his MALS from the University of Denver and his PhD from the University of Colorado. He has served as a consultant for the U.S. Information Agency in Paraguay, Poland, and Serbia; the World Bank in Paraguay; and the Chief of Staff U.S. Army Training and Leader Development Panel. For more than 10 years, he co-organized a series of Eastern European international conferences on library development in a networked world. His awards include serving twice as a Fulbright Senior Specialist in Bulgaria and Serbia, the St. Kliment Ohridski University of Sofia Blue Ribbon Medal, and the Roe R. Cross Distinguished Professor Award at Emporia State University. His research focuses on information transfer as related to knowledge management in organizations and the global information society.

KELLY VISNAK is scholarly communication librarian at the University of Wyoming, where she provides instruction, consultation, and policy support for issues related to fair use, copyright retention, open access, data management plans, and the libraries' digital scholarship repository. Previously, she served for more than a decade as a regional administrator for Emporia State University School of Library and Information Management as director of the Colorado MLS Distance Education Program. She earned her PhD from Dominican University's Library and Information Science Program. She has published in several professional journals.